FARMING OF VEGETABLE AND FRUITS

Edited by
Arit Efretuei

Arcler Press

Farming of Vegetable and Fruits

Edited by Arit Efretuei

Edition 2018

ArclerPress

Published by
Arcler Press LLC
708 3rd Avenue, 6th Floor
New York
NY 10017
United States of America
www.arclerpress.com

ISBN: 978-1-68094-564-5 (Hardback)
Printed in United States of America
Library of Congress Control Number: 2017933134

About the Editor

Arit Efretuei

Arit Efretuei obtained her PhD in Agriculture from the University of Reading, UK in 2015. Her interests are in crop nutrient uptake, soil nutrient management and mitigation of soil nutrient losses to the environment. She is presently working as a freelance agronomic writer.

Table of Contents

2 FARMING OF SOLANACEAE FAMILY

3 BRASSICACEAE FARMING

4 ALLIUM

5 FRUIT FARMINGS: AN OVERVIEW

6 CULTIVATION OF AGGREGATE FRUIT

7 CULTIVATION OF MULTIPLE FRUIT

List of Abbreviations

ITS	Internal Transcribed Spacer
ORAC	Oxygen Radical Absorbance Capacity
RL	Regular Leaf
USDA-ARS	United States Department of Agriculture Agricultural Research Service

Preface

Farming of Vegetable and Fruits is the growing of vegetable and fruit crops, primarily for use as human food. In everyday usage, a vegetable is any part of a plant that is consumed by humans as food as part of a savory meal. Vegetables are important part of healthy eating and provide a source of many nutrients, including potassium, fiber, folate (folic acid) and vitamins A, E and C. Options like broccoli, spinach, tomatoes and garlic provide additional benefits, making them a super food. Potassium may help to maintain healthy blood pressure. Dietary fiber from vegetables helps reduce blood cholesterol levels and may lower risk of heart disease. Folate (folic acid) helps the body form healthy red blood cells. Fruit has been recognized as a good source of vitamins and minerals, and for their role in preventing vitamin C and vitamin A deficiencies. People who eat fruit as part of an overall healthy diet generally have a reduced risk of chronic diseases. Fruit are important sources of many nutrients, including potassium, fiber, vitamin C and folate (folic acid). Eating fruit provides health benefits. Fruits provide nutrients vital for health and maintenance of your body. The popular distinction between vegetable and fruit is difficult to uphold. In general, those plants or plant parts that are usually consumed with the main course of a meal are popularly regarded as vegetables, while those mainly used as desserts are considered fruits. This distinction is applied in this article. Thus, cucumber and tomato, botanically fruits, since they are the portion of the plant containing seeds, are commonly regarded as vegetables. The subject of fruit and nut production deals with intensive culture of perennial plants, the fruits of which have economic significance. It is one part of the broad subject of horticulture, which also encompasses vegetable growing and production of ornamentals and flowers. This text treats the principles and practices of vegetable farming.

Content Coverage

Chapters One and *Two* provide an outline of legumes and flowering plants, respectively. The Fabaceae family, known as legumes, are one of the most important plant families in both ecological and economic terms, whereas flowering plants family is an important source of food, spice, and medicine, and many plants are used as ornamentals, including the petunia and butterfly flower.

Chapters Three and *Four* reveal on Brassicaceae plants and Allium, separately. Brassicaceae is a family of plants most commonly referred to as the mustard or the cabbage family, whereras Allium is a genus of monocotyledonous flowering plants that includes the cultivated onion, garlic, scallion, shallot and leek as well as chives and hundreds of other wild species.

Chapters Five and *Six* are basics of fruit and cultivation of aggregate fruit, correspondingly. Anthropological criminology is a field of offender profiling, based on perceived links between the nature of a crime and the personality and physical appearance of the offender. Crime prevention is the attempt to reduce and deter crime and criminals. It is applied specifically to efforts made by governments to reduce crime, enforce the law, and maintain criminal justice.

Chapter Seven presents about the multiple fruits, also called collective fruits, are fruiting bodies formed from a cluster of fruiting flowers, the inflorescence. Each flower in the inflorescence produces a fruit, but these mature into a single mass in which each flower has produced a true fruit.

CHAPTER 1

Farming of Legumes

Chapter Contents

The Fabaceae family, known as legumes, are one of the most important plant families in both ecological and economic terms. Legumes help increase soil nitrogen and provide rich sources of vegetable protein for humans, livestock, and wild animals.

The Fabaceae predominate in most vegetation types of the world. Legume species are most abundant in seasonally dry, tropical forests, but they also abound in high deserts to lowland rain forests. They are notably uncommon in high alpine sites with abundant summer rains and in the southern beech forests of the Southern Hemisphere. Woody legume species are mostly confined to tropical and subtropical habitats, but herbaceous species occur from the tropics to cold temperate regions. Thus, many legume species are available for use in reforestation or revegetation projects worldwide. Their efficient ability to acquire nitrogen, phosphorus, and other essential nutrients, whether through nodulation or association with mycorrhizae (symbiotic root fungi), allows them to easily establish on abused lands where soils are eroded, leached, or acidic. Legumes can be found as pioneer species

INTRODUCTION

The Fabaceae, Leguminosae or Papilionaceae, commonly known as the legume, pea, or bean family, are a large and economically important family of flowering plants. It includes trees, shrubs, and perennial or annual herbaceous plants, which are easily recognized by their fruit (legume) and their compound, stipulated leaves. The family is widely distributed, and is the third-largest land plant family in terms of number of species, behind only the Orchidaceae and Asteraceae, with about 751 genera and some 19,000 known species. The five largest of the genera are *Astragalus* (over 3,000 species), *Acacia* (over 1000 species), *Indigofera* (around 700 species), *Crotalaria* (around 700 species) and *Mimosa* (around 500 species), which constitute about a quarter of all legume species. The ca. 19,000 known legume species amount to about 7% of flowering plant species. Fabaceae is the most common family found in tropical rainforests and in dry forests in the Americas and Africa.

Recent molecular and morphological evidence supports the fact that the Fabaceae is a single monophyletic family. This point of view has been supported not only by the degree of interrelation shown by different groups within the family compared with that found among the Leguminosae and their closest relations, but also by all the recent phylogenetic studies based on DNA sequences. These studies confirm that the Fabaceae are a monophyletic group that is closely related to the Polygalaceae, Surianaceae and Quillajaceae families and that they belong to the order Fabales.

Flowering kudzu is a fast-growing legume with a grapelike odor.

Along with the cereals, some fruits and tropical roots a number of Leguminosae have been a staple human food for millennia and their use is closely related to human evolution.

A number are important agricultural and food plants, including *Glycine max* (soybean), *Phaseolus* (beans), *Pisum sativum* (pea), *Cicer arietinum* (chickpeas), *Medicago sativa* (alfalfa), *Arachis hypogaea* (peanut), *Lathyrus odoratus* (sweet pea), *Ceratonia siliqua* (carob), and *Glycyrrhiza glabra* (liquorice). A number of species are also weedy pests in different parts of the world, including: *Cytisus scoparius* (broom), *Robinia pseudoacacia* (black locust), *Ulex europaeus* (gorse), *Pueraria lobata* (kudzu), and a number of *Lupinus* species.

ETYMOLOGY

The name 'Fabaceae' comes from the defunct genus *Faba*, now included in *Vicia*. The term "faba" comes from Latin, and appears to simply mean "bean". Leguminosae is an older name still considered valid, and refers to the fruit of these plants, which are called legumes.

DESCRIPTION

Fabaceae range in habit from giant trees (like *Koompassia excelsa*) to small annual herbs, with the majority being herbaceous perennials. Plants have indeterminate inflorescences, which are sometimes reduced to a single flower. The flowers have a short hypanthium and a single carpel with a short gynophore, and after fertilization produce fruits that are legumes.

The fruit of *Gymnocladus dioicus*

Growth Habit

The Leguminosae have a wide variety of growth forms including trees, shrubs or herbaceous plants or even vines or lianas. The herbaceous plants can be annuals, biennials or perennials, without basal or terminal leaf aggregations. They are upright plants, epiphytes or vines. The latter support themselves by means of shoots that twist around a support or through cauline or foliar tendrils. Plants can be heliophytes, mesophytes or xerophytes.

Leaves

The leaves are usually alternate and compound. Most often they are even- or odd-pinnately compound (e.g. *Caragana* and *Robinia* respectively), often trifoliate (e.g. *Trifolium*, *Medicago*) and rarely palmately compound (e.g. *Lupinus*), in the Mimosoideae and the Caesalpinioideae commonly bipinnate (e.g. *Acacia*, *Mimosa*). They always have stipules, which can be leaf-like (e.g. *Pisum*), thorn-like (e.g. *Robinia*) or be rather inconspicuous. Leaf margins are entire or, occasionally, serrate. Both the leaves and the leaflets often have wrinkled pulvini to permit nastic movements. In some species, leaflets have evolved into tendrils (e.g. *Vicia*).

Many species have leaves with structures that attract ants that protect the plant from herbivore insects (a form of mutualism). Extrafloral nectaries are common among the Mimosoideae and the Caesalpinioideae, and are also found in some Faboideae (e.g. *Vicia sativa*). In some *Acacia*, the modified hollow stipules are inhabited by ants and are known as domatia.

Roots

Many Fabaceae host bacteria in their roots within structures called root nodules. These bacteria, known as rhizobia, have the ability to take nitrogen gas (N_2) out of the air and convert it to a form of nitrogen that is usable to the host plant (NO_3^- or NH_3). This process is called nitrogen fixation. The legume, acting as a host, and rhizobia, acting as a provider of usable nitrate, form a symbiotic relationship.

Flowers

The flowers often have five generally fused sepals and five free petals. They are generally hermaphrodite, and have a short hypanthium, usually cup shaped. There are normally ten stamens and one elongated superior ovary, with a curved style. They are usually arranged in indeterminate inflorescences. Fabaceae are typically entomophilous plants (i.e. they are pollinated by insects), and the flowers are usually showy to attract pollinators.

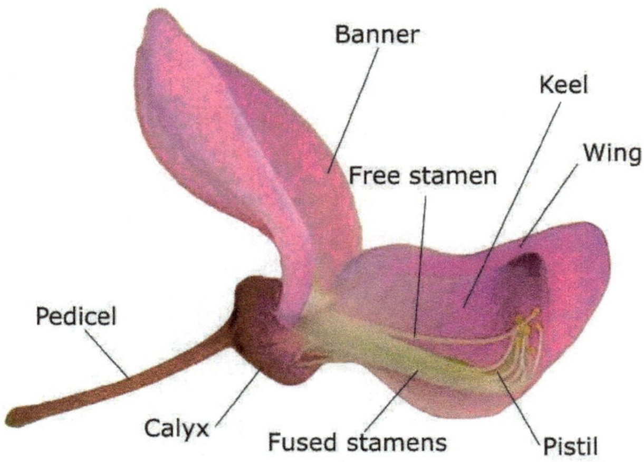

A flower of *Wisteria sinensis*, Faboideae. Two petals have been removed to show stamens and pistil

In the Caesalpinioideae, the flowers are often zygomorphic, as in *Cercis*, or nearly symmetrical with five equal petals in *Bauhinia*. The upper petal is the innermost one, unlike in the Faboideae. Some species, like some in the genus *Senna*, have asymmetric flowers, with one of the lower petals larger than the opposing one, and the style bent to one side. The calyx, corolla, or stamens can be showy in this group.

In the Mimosoideae, the flowers are actinomorphic and arranged in globose inflorescences. The petals are small and the stamens, which can be more than just 10, have long, colored filaments, which are the showiest part of the flower. All of the flowers in an inflorescence open at once.

In the Faboideae, the flowers are zygomorphic, and have a specialized structure. The upper petal, called the banner, is large and envelops the rest of the petals in bud, often reflexing when the flower blooms. The two adjacent petals, the wings, surround the two bottom petals. The two bottom petals are fused together at the apex (remaining free at the base), forming a boat-like structure called the keel. The stamens are always ten in number, and their filaments can be fused in various configurations, often in a group of nine stamens plus one separate stamen. Various genes in the *CYCLOIDEA (CYC)/ DICHOTOMA (DICH)* family are expressed in the upper (also called dorsal or adaxial) petal; in some species, such as *Cadia*, these genes are expressed throughout the flower, producing a radially symmetrical flower.

Fruit

The ovary most typically develops into a legume. A legume is a simple dry fruit that usually dehisces (opens along a seam) on two sides. A common name for this type of fruit is a "pod", although that can also be applied to a few other fruit types. A few species have evolved samarae, loments, follicles, indehiscent legumes, achenes, drupes, and berries from the basic legume fruit.

Legume of *Vicia angustifolia*

Physiology and Biochemistry

The Leguminosae are rarely cyanogenic, however, where they are, the cyanogenic compounds are derived from tyrosine, phenylalanine or leucine. They frequently contain alkaloids. Proanthocyanidins can be present either as cyanidin or delphinidine or both at the same time. Flavonoids such as kaempferol, quercitin and myricetin are often present. Ellagic acid has never been found in any of the genera or species analysed. Sugars are transported within the plants in the form of sucrose. C3 photosynthesis has been found in a wide variety of genera. The family has also evolved a unique chemistry. Pterocarpans are a class of molecules (derivatives of isoflavonoids) found only in the Fabaceae.

Ecology

Distribution and Habitat

The Fabaceae have an essentially worldwide distribution, being found everywhere except Antarctica and the high arctic. The trees are often found in tropical regions, while the herbaceous plants and shrubs are predominant outside the tropics.

Biological Nitrogen Fixation

Biological nitrogen fixation (BNF, performed by the organisms called diazotrophs) is a very old process that probably originated in the Archean eon when the primitive atmosphere lacked oxygen. It is only carried out by Euryarchaeota and just 6 of the more than 50 phyla of bacteria. Some of these lineages co-evolved together with the flowering plants establishing the molecular basis of a mutually beneficial symbiotic relationship. BNF is carried out in nodules that are mainly located in the root cortex, although they are occasionally located in the stem as in *Sesbania rostrata*. The spermatophytes that co-evolved with actinorhizal diazotrophs (*Frankia*) or with rhizobia to establish their symbiotic relationship belong to 11 families contained within the Rosidae clade (as established by the gene molecular phylogeny of *rbcL*, a gene coding for part of the RuBisCO enzyme in the chloroplast).

Roots of *Vicia* with white root nodules visible.

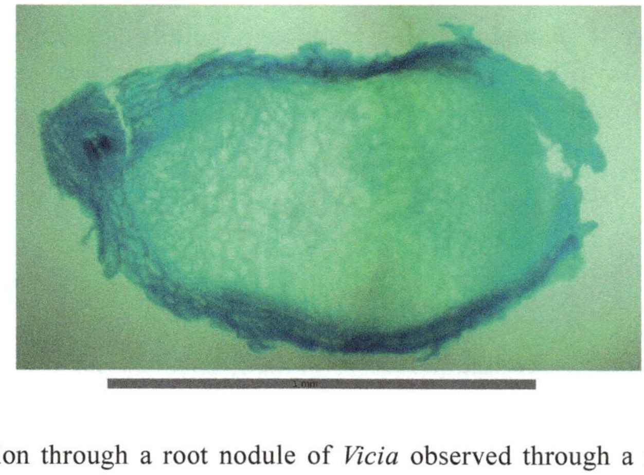

Cross-section through a root nodule of *Vicia* observed through a microscope.

This grouping indicates that the predisposition for forming nodules probably only arose once in flowering plants and that it can be considered as an ancestral characteristic that has been conserved or lost in certain lineages. However, such a wide distribution of families and genera within this lineage indicates that nodulation had multiple origins. Of the 10 families within the Rosidae, 8 have nodules formed by actinomyces (Betulaceae, Casuarinaceae, Coriariaceae, Datiscaceae, Elaeagnaceae, Myricaceae, Rhamnaceae and Rosaceae), and the two remaining families, Ulmaceae and Fabaceae have nodules formed by rhizobia.

The rhizobia and their hosts must be able to recognize each other for nodule formation to commence. Rhizobia are specific to particular host species although a rhizobia species may often infect more than one host species. This means that one plant species may be infected by more than one species of bacteria. For example, nodules in *Acacia senegal* can contain seven species of rhizobia belonging to three different genera. The most distinctive characteristics that allow rhizobia to be distinguished apart are the rapidity of their growth and the type of root nodule that they form with their host. Root nodules can be classified as being either indeterminate, cylindrical and often branched, and determinate, spherical with prominent lenticels. Indeterminate nodules are characteristic of legumes from temperate climates, while determinate nodules are commonly found in species from tropical or subtropical climates.

Nodule formation is common throughout the leguminosae, it is found in the majority of its members that only form an association with rhizobia, which in turn form an exclusive symbiosis with the leguminosae (with the exception of *Parasponia*, the only genus of the 18 Ulmaceae genera that is capable of forming nodules). Nodule formation is present in all the leguminosae sub-families, although it is less common in the Caesalpinioideae. All types of nodule formation are present in the sub-family Papilionoideae: indeterminate (with the meristem retained), determinate (without meristem) and the type included in *Aeschynomene*. The latter two are thought to be the most modern and specialised type of nodule as they are only present in some lines of the Papilionoideae sub-family. Even though nodule formation is common in the two monophyletic subfamilies Papilionoideae and Mimosoideae they also contain species that do not form nodules. The presence or absence of nodule-forming species within the three sub-families indicates that nodule formation has arisen several times during the evolution of the leguminosae and that this ability has been lost in some lineages. For example, within the genus *Acacia*, a member of the Mimosoideae, *A. pentagona* does not form nodules, while other species of the same genus readily form nodules, as is the case for *Acacia senegal*, which forms both rapidly and slow growing rhizobial nodules.

EVOLUTION, PHYLOGENY AND TAXONOMY

Evolution

The order Fabales contains around 7.3% of eudicot species and the greatest part of this diversity is contained in just one of the four families that order contains: Fabaceae. This clade also includes the Polygalaceae, Surianaceae and Quillajaceae families and its origins date back 94 to 89 million years, although it started its diversification some 79 to 74 million years ago. In fact, the Fabaceae have diversified during the early tertiary to become a ubiquitous part of the modern earth's biota, along with many other families belonging to the flowering plants.

The Fabaceae have an abundant and diverse fossil record, especially for the Tertiary period. Fossils of flowers, fruit, leaves, wood and pollen from this period have been found in numerous locations. The earliest fossils that can be definitively assigned to the Fabaceae appeared in the late Palaeocene (approximately 56 million years ago). Representatives of the 3 sub-families traditionally recognised as being members of the Fabaceae – Cesalpinioideae, Papilionoideae and Mimosoideae — as well as members of the large clades within these sub-families – such as the genistoides – have been found in periods a little later, starting between 55 and 50 million years ago. In fact, a wide variety of taxa representing the main lineages in the Fabaceae have been found in the fossil record dating from the middle to the late Eocene, suggesting that the majority of the modern Fabaceae groups were already present and that a broad diversification occurred during this period. Therefore, the Fabaceae started their diversification approximately 60 million years ago and the most important clades separated some 50 million years ago.The age of the main Cesalpinioideae clades have been estimated as between 56 and 34 million years and the basal group of the Mimosoideae as 44 ± 2.6 million years. The division between Mimosoideae and Faboideae is dated as occurring between 59 and 34 million years ago and the basal group of the Faboideae as 58.6 ± 0.2 million years ago. It has been possible to date the divergence of some of the groups within the Faboideae, even though diversification within each genus was relatively recent. For instance, *Astragalus* separated from the *Oxytropis* some 16 to 12 million years ago. In addition, the separation of the aneuploid species of *Neoastragalus* started 4 million years ago. *Inga,* another genus of the Papilionoideae with approximately 350 species, seems to have diverged in the last 2 million years.

It has been suggested, based on fossil and phylogenetic evidence, that legumes originally evolved in arid and/or semi-arid regions along the Tethys seaway during the Palaeogene Period. However, others contend that Africa (or even the Americas) cannot yet be ruled out as the origin of the family.

The current hypothesis about the evolution of the genes needed for nodulation is that they were recruited from other pathways after a polyploidy event. Several different pathways have been implicated as donating duplicated genes to the pathways need for nodulation. The main donors to the pathway were the genes associated with the arbuscular mycorrhiza symbiosis genes, the pollen tube formation genes and the haemoglobin genes. One of the main genes shown to be shared between the arbuscular mycorrhiza pathway and the nodulation pathway is SYMRK and it is involved in the plant-bacterial recognition. The pollen tube growth is similar to the infection thread development in that infection threads grow in a polar manner that is similar to a pollen tubes polar growth towards the ovules. Both pathways include the same type of enzymes, pectin-degrading cell wall enzymes. The enzymes needed to reduce nitrogen, nitrogenases, require a substantial input of ATP but at the same time are sensitive to free oxygen. To meet the requirements of this paradoxical situation, the plants express a type of haemoglobin called leghaemoglobin that is believed to be recruited after a duplication event. These three genetic pathways are believed to be part of a gene duplication event then recruited to work in nodulation.

Phylogeny and Taxonomy

Phylogeny

The phylogeny of the legumes has been the object of many studies by research groups from around the world. These studies have used morphology, DNA data (the chloroplast intron *trnL*, the chloroplast genes *rbcL* and *matK*, or the ribosomal spacers *ITS*) and cladistic analysis in order to investigate the relationships between the family's different lineages. The studies have confirmed that the traditional sub-families Mimosoideae and Papilionoideae are each monophyletic but both are nested within the paraphyletic sub-family Caesalpinioideae. All the different approaches have yielded similar results regarding the relationships between the family's main clades, as shown in the cladogram below.

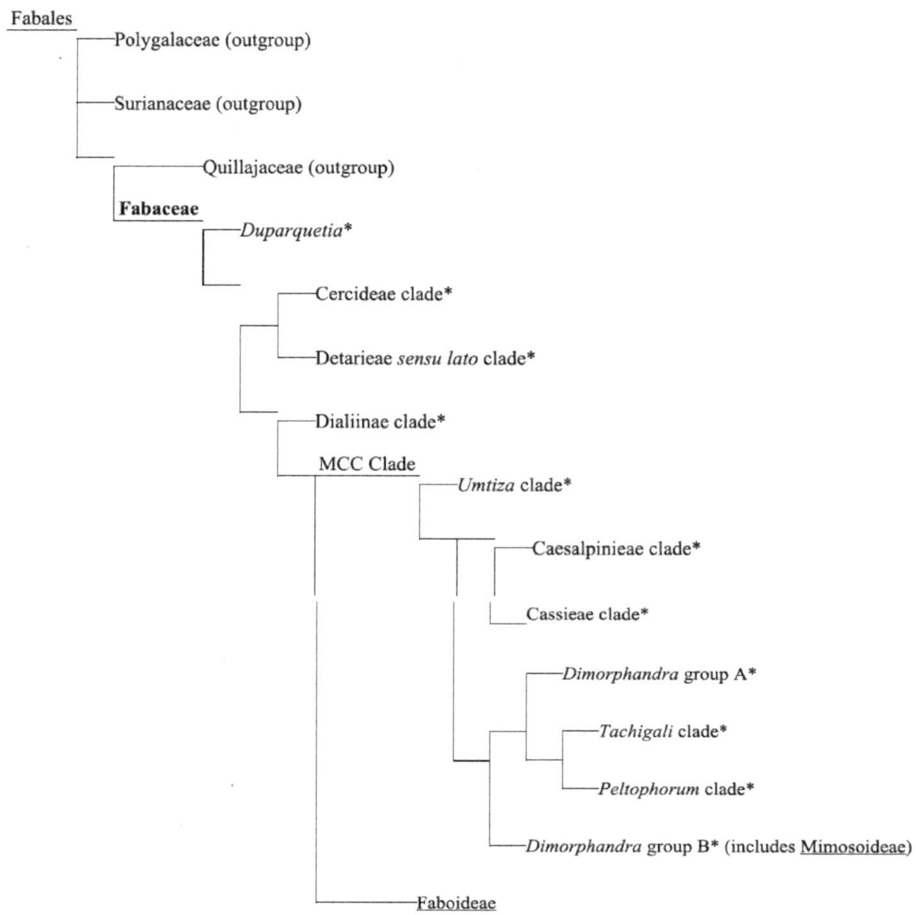

Asterisks (*) indicate clades traditionally assigned to Caesalpinioideae; the other two subfamilies (which are nested within Caesalpinioideae) are underlined.

Taxonomy

The Fabaceae are placed in the order Fabales according to most taxonomic systems, including the APG III system. The family includes three subfamilies:

- Mimosoideae: 80 genera and 3,200 species. Mostly tropical and warm temperate Asia and America. *Mimosa*, *Acacia*.
- Caesalpinioideae: 170 genera and 2,000 species, cosmopolitan. *Caesalpinia*, *Senna*, *Bauhinia*, *Amherstia*.

Faboideae (Papilionoideae): 470 genera and 14,000 species, cosmopolitan. *Astragalus*, *Lupinus*.

These three subfamilies have been alternatively treated at the family level, as in the Cronquist and Dahlgren systems. However, this choice has not been supported by late 20th and early 21st century evidence, which has shown the Caesalpinioideae to be paraphyletic and the Fabaceae *sensu lato* to be monophyletic. While the Mimosoideae and the Faboideae are largely monophyletic, the Caesalpinioideae appear to be paraphyletic and the tribe Cercideae is probably sister to the rest of the family. Moreover, there are a number of genera whose placement into the Caesalpinioideae is not always agreed on (e.g. *Dimorphandra*).

Genera

The 730 genera included in this family can be viewed on the following three pages:

- List of Mimosoideae genera
- List of Caesalpinioideae genera
- List of Faboideae genera

Economic and Cultural Importance

Legumes are economically and culturally important plants due to their extraordinary diversity and abundance, the wide variety of edible vegetables they represent and due to the variety of uses they can be put to: in horticulture and agriculture, as a food, for the compounds they contain that have medicinal uses and for the oil and fats they contain that have a variety of uses.

Food and Forage

The history of legumes is tied in closely with that of human civilization, appearing early in Asia, the Americas (the common bean, several varieties) and Europe (broad beans) by 6,000 BCE, where they became a staple, essential as a source of protein.

Their ability to fix atmospheric nitrogen reduces fertilizer costs for farmers and gardeners who grow legumes, and means that legumes can be used in a crop rotation to replenish soil that has been depleted of nitrogen. Legume seeds and foliage have a comparatively higher protein content than non-legume materials, due to the additional nitrogen that legumes receive through the process. Some legume species perform hydraulic lift, which makes them ideal for intercropping.

Farmed legumes can belong to numerous classes, including forage, grain, blooms, pharmaceutical/industrial, fallow/green manure and timber species, with most commercially farmed species filling two or more roles simultaneously.

There are of two broad types of forage legumes. Some, like alfalfa, clover, vetch, and *Arachis*, are sown in pasture and grazed by livestock. Other forage legumes such as *Leucaena* or *Albizia* are woody shrub or tree species that are either broken down by livestock or regularly cut by humans to provide stock feed.

Grain legumes are cultivated for their seeds, and are also called pulses. The seeds are used for human and animal consumption or for the production of oils for industrial uses. Grain legumes include both herbaceous plants like beans, lentils, lupins, peas and peanuts. and trees such as carob, mesquite and tamarind.

Bloom legume species include species such as lupin, which are farmed commercially for their blooms as well as being popular in gardens worldwide. *Laburnum*, *Robinia*, *Gleditsia*, *Acacia*, *Mimosa*, and *Delonix* are ornamental trees and shrubs.

Industrial farmed legumes include *Indigofera*, cultivated for the production of indigo, *Acacia*, for gum arabic, and *Derris*, for the insecticide action of rotenone, a compound it produces.

Fallow or green manure legume species are cultivated to be tilled back into the soil to exploit the high nitrogen levels found in most legumes. Numerous legumes are farmed for this purpose, including *Leucaena*, *Cyamopsis* and *Sesbania*.

Various legume species are farmed for timber production worldwide, including numerous *Acacia* species, *Dalbergia* species, and *Castanospermum australe*.

Melliferous plants offer nectar to bees and other insects to encourage them to carry pollen from the flowers of one plant to others thereby ensuring pollination.A number of legume species are good nectar providers such as alfalfa, white clover, sweet clover and various Prosopis species. Many plants in the *Fabaceae* family are an important source of pollen for the bumblebee species *Bombus hortorum*. This bee species is especially fond of one species in particular; *Trifolium pratense*, also known as red clover, is a popular food source in the diet of *Bombus hortorum*.

Industrial Uses

Natural Gums

Natural gums are vegetable exudates that are released as the result of damage to the plant such as that resulting from the attack of an insect or a natural or artificial cut. These exudates contain heterogeneous polysaccharides formed of different sugars and usually containing uronic acids. They form viscous colloidal solutions. There are different species that produce gums. The most important of these species belong to the leguminosae. They are widely used in the pharmaceutical, cosmetic, food and textile sectors. They also have interesting therapeutic properties; for example gum arabic is antitussive and anti-inflammatory. The most well known gums are tragacanth (*Astragalus gummifer*), gum arabic (*Acacia senegal*) and guar gum (*Cyamopsis tetragonoloba*).

Dyes

The species used to produce dyes include the following: Logwood *Haematoxylon campechianum*; a large spiny tree that can grow up to 15 m tall. Its cork is thin and soft and its wood is hard. The heartwood is used to produce dyes that are red and purple. The histological stain called haematoxylin is produced from this species.

Indigo colorant

Brazilwood tree (*Caesalpinia echinata*) is similar to the previous tree but smaller and with red or purple flowers. The wood is also used to produce a red or purple dye. The Madras thorn (*Pithecallobium dulce*) is another spiny tree native to Latin America, it grows up to 4 m high and has yellow or green flowers that grow in florets. Its fruit is reddish and is used to produce a yellow dye. Indigo dye is extracted from the True indigo plant *Indigofera tinctoria* that is native to Asia. In Central and South America dyes are produced from two species related to this species, indigo from *Indigofera suffruticosa* and Natal indigo from *Indigofera arrecta*.yellow dye is extracted from *Butea monosperma* commonly called as flame of the forest.

Ornamentals

Legumes have been used as ornamental plants throughout the world for many centuries. Their vast diversity of heights, shapes, foliage and flower colour means that this family is commonly used in the design and planting of everything from small gardens to large parks. The following is a list of the main ornamental legume species, listed by sub-family.

- Subfamily Caesalpinioideae: *Bauhinia forficata*, *Caesalpinia gilliesii*, *Caesalpinia spinosa*, *Ceratonia siliqua*, *Cercis siliquastrum*, *Gleditsia triacanthos*, *Gymnocladus dioica*, *Parkinsonia aculeata*, *Senna multiglandulosa*.

- Subfamily Mimosoideae: *Acacia caven*, *Acacia cultriformis*, *Acacia dealbata*, *Acacia karroo*, *Acacia longifolia*, *Acacia melanoxylon*, *Acacia paradoxa*, *Acacia retinodes*, *Acacia saligna*, *Acacia verticillata*, *Acacia visco*, *Albizzia julibrissin*, *Calliandra tweediei*, *Paraserianthes lophantha*, *Prosopis chilensis*.
- Subfamily Faboideae: *Clianthus puniceus*, *Citysus scoparius*, *Erythrina crista-galli*, *Erythrina falcata*, *Laburnum anagyroides*, *Lotus peliorhynchus*, *Lupinus arboreus*, *Lupinus polyphyllus*, *Otholobium glandulosum*, *Retama monosperma*, *Robinia hispida*, *Robinia luxurians*, *Robinia pseudoacacia*, *Sophora japonica*, *Sophora macnabiana*, *Sophora macrocarpa*, *Spartium junceum*, *Teline monspessulana*, *Tipuana tipu*, *Wisteria sinensis*.

The Cockspur Coral Tree *Erythrina crista-galli* is one of many leguminosae used as ornamental plants. In addition, it is the National Flower of Argentina and Uruguay.

Emblematic Leguminosae

- The Cockspur Coral Tree (*Erythrina crista-galli*), is the National Flower of Argentina and Uruguay.
- The Elephant ear tree (*Enterolobium cyclocarpum*) is the national tree of Costa Rica, by Executive Order of 31 August 1959.
- The Brazilwood tree (*Caesalpinia echinata*) has been the national tree of Brazil since 1978.
- The Golden wattle *Acacia pycnantha* is Australia's national flower.
- The Hong Kong Orchid tree *Bauhinia blakeana* is the national flower of Hong Kong.

Image Gallery

Acacia baileyana (Wattle)

Loments of *Alysicarpus vaginalis*

Calliandra emarginata

Desmodium gangeticum

Dichrostachys cinerea Sickle Bush

Indigofera gerardiana

Tendrils of *Lathyrus odoratus* (Sweet pea)

Inflorescence of *Lupinus arboreus* (Yellow bush lupin)

Pisum sativum (Peas); note the leaf-like stipules

Smithia conferta

Trifolium repens in Kullu District of Himachal Pradesh, India.

Kashubian vetch – Kashubia

Zornia gibbosa

PEA

The pea is a small spherical seed of the pod fruit Pisum sativum seed or the seed-pod of the pod fruit *Pisum sativum*.Each pod contains several peas. Pea pods are fruits since they contain seeds and they develop from the ovary of a (pea) flower. The name is also used to describe other edible seeds from the Fabaceae such as the pigeon pea (*Cajanus cajan*), the cowpea (*Vigna unguiculata*), and the seeds from several species of *Lathyrus*.

P. sativum is an annual plant, with a life cycle of one year. It is a cool-season crop grown in many parts of the world; planting can take place from winter to early summer depending on location. The average pea weighs between 0.1 and 0.36 grams. The immature peas (and in snow peas the tender pod as well) are used as a vegetable, fresh, frozen or canned; varieties of the species typically called field peas are grown to produce dry peas like the split pea shelled from the matured pod. These are the basis of pease porridge and pea soup, staples of medieval cuisine; in Europe, consuming fresh immature green peas was an innovation of Early Modern cuisine.

Peas are contained within a pod

Pea plant: Pisum sativum

The wild pea is restricted to the Mediterranean basin and the Near East. The earliest archaeological finds of peas date from the late neolithic era of current Greece, Syria, Turkey and Jordan. In Egypt, early finds date from *ca.* 4800–4400 BC in the Nile delta area, and from *ca.* 3800–3600 BC in Upper Egypt. The pea was also present in Georgia in the 5th millennium BC. Farther east, the finds are younger. Peas were present in Afghanistan *ca.* 2000 BC, in Harappa, Pakistan, and in northwest India in 2250–1750 BC. In the second half of the 2nd millennium BC, this pulse crop appears in the Ganges Basin and southern India.

Description

A pea is a most commonly green, occasionally golden yellow, or infrequently purple pod-shaped vegetable, widely grown as a cool season vegetable crop. The seeds may be planted as soon as the soil temperature reaches 10 °C (50 °F), with the plants growing best at temperatures of 13 to 18 °C (55 to 64 °F). They do not thrive in the summer heat of warmer temperate and lowland tropical climates, but do grow well in cooler, high altitude, tropical areas. Many cultivars reach maturity about 60 days after planting.

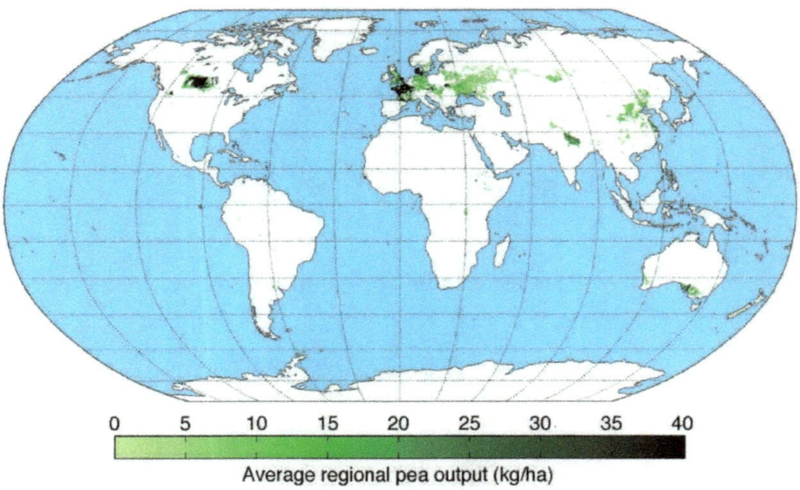

Worldwide pea yield

Peas, green, raw	
Nutritional value per 100 g (3.5 oz)	
Energy	339 kJ (81 kcal)
Carbohydrates	14.45 g

Sugars	5.67 g
Dietary fiber	5.1 g
Fat	0.4 g
Protein	5.42 g
Vitamins	
Vitamin A equiv. beta-carotene lutein zeaxanthin	(5%) 38 μg (4%) 449 μg 2477 μg
Thiamine (B1)	(23%) 0.266 mg
Riboflavin (B2)	(11%) 0.132 mg
Niacin (B3)	(14%) 2.09 mg
Vitamin B6	(13%) 0.169 mg
Folate (B9)	(16%) 65 μg
Vitamin C	(48%) 40 mg
Vitamin E	(1%) 0.13 mg
Vitamin K	(24%) 24.8 μg
Minerals	
Calcium	(3%) 25 mg

Iron	(11%) 1.47 mg
Magnesium	(9%) 33 mg
Manganese	(20%) 0.41 mg
Phosphorus	(15%) 108 mg
Potassium	(5%) 244 mg
Sodium	(0%) 5 mg
Zinc	(13%) 1.24 mg
Link to USDA Database entry	
• Units • μg = micrograms • mg = milligrams • IU = International units	
Percentages are roughly approximated using US recommendations for adults. Source: USDA Nutrient Database	

Peas have both low-growing and vining cultivars. The vining cultivars grow thin tendrils from leaves that coil around any available support and can climb to be 1–2 m high. A traditional approach to supporting climbing peas is to thrust branches pruned from trees or other woody plants upright into the soil, providing a lattice for the peas to climb. Branches used in this fashion are sometimes called **pea brush**. Metal fences, twine, or netting supported by a frame are used for the same purpose. In dense plantings, peas give each other some measure of mutual support. Pea plants can self-pollinate.

History

In early times, peas were grown mostly for their dry seeds. From plants growing wild in the Mediterranean basin, constant selection since the Neolithic dawn of agriculture improved their yield. In the early 3rd century BC Theophrastus mentions peas among the pulses that are sown late in the winter because of their tenderness. In the first century AD Columella mentions them in *De re rustica*, when Roman legionaries still gathered wild peas from the sandy soils of Numidia and Judea to supplement their rations.

Pea in a painting by Mateusz Tokarski, ca. 1795 (National Museum in Warsaw)

In the Middle Ages, field peas are constantly mentioned, as they were the staple that kept famine at bay, as Charles the Good, count of Flanders, noted explicitly in 1124. In the 13th century the poet Guillaume de Villeneuve noted

J'ay pois en cosse touz noviaux among the street cries of Paris.

Green "garden" peas, eaten immature and fresh, were an innovative luxury of Early Modern Europe. In England, the distinction between "field peas" and "garden peas" dates from the early 17th century: John Gerard and John Parkinson both mention garden peas. Sugar peas, which the French soon called *mange-tout*, for they were consumed pods and all, were introduced to France from the market gardens of Holland in the time of Henri IV, through the French ambassador. Green peas were introduced from Genoa to the court of Louis XIV of France in January 1660, with some staged fanfare; a hamper of them were presented before the King, and then were shelled by the Sovoyan comte de Soissons, who had married a niece of Cardinal Mazarin; little dishes of peas were then presented to the King, the Queen, Cardinal Mazarin and Monsieur, the king's brother. Immediately established and grown for earliness warmed with manure and protected under glass, they were still a luxurious delicacy in 1696, when Mme de Maintenon and Mme de Sevigné each reported that they were "a fashion, a fury."

Modern split peas, with their indigestible skins rubbed off, are a development of the later 19th century.

Modern Culinary Use

In modern times peas are usually boiled or steamed, which breaks down the cell walls and makes the taste sweeter and the nutrients more bioavailable. Along with broad beans and lentils, these formed an important part of the diet of most people in the Middle East, North Africa and Europe during the Middle Ages. By the 17th and 18th centuries, it had become popular to eat peas "green", that is, while they are immature and right after they are picked.

Split peas (raw)	
 Yellow split peas	
Nutritional value per 100 g (3.5 oz)	
Energy	1,425 kJ (341 kcal)
Carbohydrates	60 g
Sugars	8 g
Dietary fiber	26 g
Fat	1 g
Protein	25 g
Vitamins	
Thiamine (B1)	(61%) 0.7 mg
Pantothenic acid (B5)	(34%) 1.7 mg
Folate (B9)	(69%) 274 μg
Minerals	
Iron	(31%) 4 mg

• Units • μg = micrograms • mg = milligrams • IU = International units
Percentages are roughly approximated using US recommendations for adults. Source: USDA Nutrient Database

Fresh peas for sale in their pods on a UK market stall

Frozen green peas

A basket of peas in pods

New cultivars of peas were developed by the English during this time, which became known as "garden" or "English" peas. The popularity of green peas spread to North America. Thomas Jefferson grew more than 30 cultivars of peas on his estate. With the invention of canning and freezing of foods, green peas became available year-round, and not just in the spring as before.

Peas in fried rice

Fresh peas are often eaten boiled and flavored with butter and/or spearmint as a side dish vegetable. Salt and pepper are also commonly added to peas when served. Fresh peas are also used in pot pies, salads and casseroles. Pod peas (particularly sweet cultivars called *mange tout* and "sugar peas", or the flatter "snow peas," called *hé lán dòu*, in Chinese) are used in stir-fried dishes, particularly those in American Chinese cuisine. Pea pods do not keep well once picked, and if not used quickly, are best preserved by drying, canning or freezing within a few hours of harvest.

In India, fresh peas are used in various dishes such as *aloo matar* (curried potatoes with peas) or *matar paneer* (paneer cheese with peas), though they can be substituted with frozen peas as well. Peas are also eaten raw, as they are sweet when fresh off the bush. Split peas are also used to make *dhal*, particularly in Guyana, and Trinidad, where there is a significant population of Indians.

Dried peas are often made into a soup or simply eaten on their own. In Japan, China, Taiwan and some Southeast Asian countries, including Thailand, the Philippines and Malaysia, peas are roasted and salted, and eaten as snacks. In the Philippines, peas, while still in their pods, are a common ingredient in viands and pansit. In the UK, dried yellow split peas are used to make pease pudding (or "pease porridge"), a traditional dish. In North America, a similarly traditional dish is split pea soup.

Pea soup is eaten in many other parts of the world, including northern Europe, parts of middle Europe, Russia, Iran, Iraq and India. In Sweden it is called *ärtsoppa*, and is eaten as a traditional Swedish food which predates the Viking age. This food was made from a fast-growing pea that would mature in a short growing season. *Ärtsoppa* was especially popular among the poor, who traditionally only had one pot and everything was cooked together for a dinner using a tripod to hold the pot over the fire.

In Chinese cuisine, the tender new growth [leaves and stem] are commonly used in stir-fries. Much like picking the leaves for tea, the farmers pick the tips off of the pea plant.

In Greece, Tunisia, Turkey, Cyprus, and other parts of the Mediterranean, peas are made into a stew with lamb and potatoes.

In Hungary and Serbia, pea soup is often served with dumplings and spiced with hot paprika.

In the United Kingdom, dried, rehydrated and mashed marrowfat peas, known by the public as mushy peas, are popular, originally in the north of England, but now ubiquitously, and especially as an accompaniment to fish and chips or meat pies, particularly in fish and chip shops. Sodium bicarbonate is sometimes added to soften the peas. In 2005, a poll of 2,000 people revealed the pea to be Britain›s seventh favorite culinary vegetable.

Processed peas are mature peas which have been dried, soaked and then heat treated (processed) to prevent spoilage—in the same manner as pasteurizing. Cooked peas are sometimes sold dried and coated with wasabi, salt, or other spices.

In North America pea milk is produced and sold as an alternative to cow milk for a variety of reasons.

Manufacturing Frozen Peas

In order to freeze and preserve peas, they must first be grown, picked, and shelled. Usually, the more tender the peas are, the more likely that they will be used in the final product. The peas must be put through the process of freezing shortly after being picked so that they do not spoil too soon. Once the peas have been selected, they are placed in ice water and allowed to cool. After, they are sprayed with water to remove any residual dirt or dust that may remain on them. The next step is blanching. The peas are boiled for a few minutes to remove any enzymes that may shorten their shelf life. They are then cooled and removed from the water. The final step is the actual freezing to produce the final product. This step may vary considerably; some companies freeze their peas by air blast freezing, where the vegetables are put through a tunnel at high speeds and frozen by cold air. Finally, the peas are packaged and shipped out for retail.

Grading

Pea grading involves sorting peas by size, in which the smallest peas are graded as the highest quality for their tenderness. Brines may be used, in which peas are floated, from which their density can be determined.

Nutritional Value

Peas are starchy, but high in fiber, protein, vitamin A, vitamin B6, vitamin C, vitamin K, phosphorus, magnesium, copper, iron, zinc and lutein. Dry weight is about one-quarter protein and one-quarter sugar. Pea seed peptide fractions have less ability to scavenge free radicals than glutathione, but greater ability to chelate metals and inhibit linoleic acid oxidation.

Varieties

There are many varieties (cultivars) of garden peas. Some of the most common varieties are listed here. *PMR* indicates some degree of powdery mildew resistance; *afila* types, also called semi-leafless, have clusters of tendrils instead of leaves. Unless otherwise noted these are so called dwarf varieties which grow to an average height of about 1m. Extra dwarf are suitable for container growing, reaching only about 25 cm. Semi-tall reaches about 1.5m and tall grows to about 2m.

- Alaska, 55 days (smooth seeded)
- Tom Thumb / Half Pint, 55 days (heirloom, extra dwarf)
- Thomas Laxton (heirloom) / Laxton's Progress / Progress #9, 60–65 days
- Mr. Big, 60 days, 2000 AAS winner
- Little Marvel, 63 days, 1934 AAS winner
- Early Perfection, 65 days
- Kelvedon Wonder, 65 days, 1997 RHS AGM winner
- Sabre, 65 days, PMR
- Homesteader / Lincoln, 67 days (heirloom, known as *Greenfeast* in Australia and New Zealand)

- Miragreen, 68 days (semi-tall climber)
- Serge, 68 days, PMR, afila
- Wando, 68 days
- Green Arrow, 70 days
- Recruit, 70 days, PMR, afila
- Tall Telephone / Alderman, 75 days (heirloom, tall climber)

Other variations of *P. sativum* include:

- *Pisum sativum* var. *macrocarpon* is commonly known as the snow pea.
- *Pisum sativum* var. *macrocarpon* is known as the sugar snap pea or snap pea.

Both of these are eaten whole before the pod reaches maturity and are hence also known as *mange-tout*, French for "eat all". The snow pea pod is eaten flat, while in sugar/snap peas, the pod becomes cylindrical, but is eaten while still crisp, before the seeds inside develop.

Usually 4-5 pods are grown from one seed.

Pests and Diseases

A variety of diseases affect peas through a number of pathogens, including insects, viruses, bacteria and fungi. In particular, virus disease of peas has worldwide economic importance.

Additionally, insects such as the pea leaf weevil (*Sitona lineatus*) can damage peas and other pod fruits. The pea leaf weevil is native to Europe, but has spread to other places such as Alberta, Canada. They are about 3.5 millimetres (0.14 in)—5.5 millimetres (0.22 in) long and are distinguishable by three light-coloured stripes running length-wise down the thorax. The weevil larvae feed on the root nodules of pea plants, which are essential to the plants' supply of nitrogen, and thus diminish leaf and stem growth. Adult weevils feed on the leaves and create a notched, "c-shaped" appearance on the outside of the leaves.

Peas in Science

In the mid-19th century, Austrian monk Gregor Mendel's observations of pea pods led to the principles of Mendelian genetics, the foundation of modern genetics. He ended up growing and examining about 28,000 pea plants in the course of his experiments. Mendel chose peas for his experiments because he could grow them easily, develop pure-bred strains, protect them from cross-pollination, and control their pollination. Mendel cross-bred tall and dwarf pea plants, green and yellow peas, purple and white flowers, wrinkled and smooth peas, and a few other traits. He then observed the resulting offspring. In each of these cases, one trait is dominant and all the offspring, or Filial-1 (abbreviated F_1) generation, showed the dominant trait.

Pea flowers

Then he crossed members of the F_1 generation together and observed their offspring, the Filial-2 (abbreviated F_2) generation. The F_2 plants had the dominant trait in approximately a 3:1 ratio. Mendel reasoned that each parent had a 'vote' in the appearance of the offspring, and the non-dominant, or recessive, trait appeared only when it was inherited from both parents. He did further experiments that showed each trait is separately inherited. Unwittingly, Mendel had solved a major problem with Charles Darwin's theory of evolution: how new traits were preserved and not blended back into the population, a question Darwin himself did not answer. Mendel's work was published in an obscure Austrian journal and was not rediscovered until about 1900.

Recently, extracts from garden pea have shown inhibitory activity on porcine pancreatic lipase in vitro.

Peas in Medicine

Some people experience allergic reactions to peas, as well as lentils, with vicilin or convicilin as the usual allergens.

Favism, or Fava-bean-ism, is a genetic deficiency of the enzyme glucose-6-phosphate dehydrogenase that affects Jews, other Middle Eastern Semitic peoples and other descendants of the Mediterranean coastal regions. In this condition, the toxic reaction to eating most, if not all, beans is hemolytic anemia, and in severe cases the released circulating free hemoglobin causes acute kidney injury.

Nitrogen-Fixing Ability

Peas, like many legumes, contain symbiotic bacteria called *Rhizobia* within root nodules of their root systems. These bacteria have the special ability of fixing nitrogen from atmospheric, molecular nitrogen (N_2) into ammonia (NH_3). The chemical reaction is:

$$N_2 + 8H^+ + 8e^- \rightarrow 2NH_3 + H_2$$

Ammonia is then converted to another form, ammonium (NH_4^+), usable by (some) plants by the following reaction:

$$NH_3 + H^+ \rightarrow NH_4^+$$

This arrangement means that the root nodules are sources of nitrogen for peas and many legumes, making them relatively rich in plant proteins. All proteins contain nitrogenous amino acids. Nitrogen is therefore a necessary ingredient in the production of proteins. Hence, peas and many legumes are among the best sources of plant protein.

When a pea plant dies in the field, for example following the harvest, all of its remaining nitrogen, incorporated into amino acids inside the remaining plant parts, is released back into the soil. In the soil, the amino acids are converted to nitrate (NO_3^-), making the nitrogen available to other plants, thereby serving as fertilizer for future crops.

Bioplastics

Bioplastics can be made using pea starch.

Etymology

The term *pea* originates from the Latin word *pisum*, which is the latinisation of the Greek *pison*, neuter of *pisos* "pea". It was adopted into English as the noun *pease* (plural *peasen*), as in pease pudding. However, by analogy with other plurals ending in *-s*, speakers began construing *pease* as a plural and constructing the singular form by dropping the *-s*, giving the term *pea*. This process is known as back-formation.

Dried peas

The name *marrowfat pea* for mature dried peas is recorded by the *Oxford English Dictionary* as early as 1733. The fact that an export cultivar popular in Japan is called *Maro* has led some people to assume mistakenly that the English name *marrowfat* is derived from Japanese.

Pea Festival

The annual 'Peasenhall Pea Festival' in the English village of Peasenhall, Suffolk attracts hundreds of visitors every year, with events such as Pea Shooting, the World Pea Podding Championships and National Pea Eating competition. In 2012, the Pea Festival had an OlymPEAn theme, celebrating the London 2012 Olympics.

BEAN

Bean is a common name for large seeds of several genera of the flowering plant family Fabaceae (also known as Leguminosae) which are used for human or animal food.

"Painted Pony" dry bean (Phaseolus vulgaris)

Bean plant

Terminology

The word "bean" and its Germanic cognates (e.g., *Bohne*) have existed in common use in West Germanic languages since before the 12th century, referring to broad beans and other pod-borne seeds. This was long before the New World genus *Phaseolus* was known in Europe. After Columbian-era contact between Europe and the Americas, use of the word was extended to pod-borne seeds of *Phaseolus*, such as the common bean and the runner bean, and the related genus *Vigna*. The term has long been applied generally to many other seeds of similar form, such as Old World soybeans, peas, chickpeas (garbanzo beans), other vetches, and lupins, and even to those with slighter resemblances, such as coffee beans, vanilla beans, castor beans, and cocoa beans. Thus the term "bean" in general usage can mean a host of different species.

Seeds called "beans" are often included among the crops called "pulses" (legumes), although a narrower prescribed sense of "pulses" reserves the word for leguminous crops harvested for their dry grain. The term *bean* usually excludes legumes with tiny seeds and which are used exclusively for forage, hay, and silage purposes (such as clover and alfalfa). According to the United Nations Food and Agriculture Organization the term *bean* should include only species of *Phaseolus*; however, enforcing that prescription has proven difficult for several reasons. One is that in the past, several species, including *Vigna angularis* (adzuki bean), *mungo* (black gram), *radiata* (green gram), and *aconitifolia* (moth bean), were classified as *Phaseolus* and later reclassified. Another is that it is not surprising that the prescription on limiting the use of the word, because it tries to replace the word's older senses with a newer one, has never been consistently followed in general usage.

Cultivation

Unlike the closely related pea, beans are a summer crop that need warm temperatures to grow. Maturity is typically 55–60 days from planting to harvest. As the bean pods mature, they turn yellow and dry up, and the beans inside change from green to their mature colour. As a vine, bean plants need external support, which may be provided in the form of special "bean cages" or poles. Native Americans customarily grew them along with corn and squash (the so-called Three Sisters), with the tall cornstalks acting as support for the beans.

Field beans (broad beans, *Vicia faba*), ready for harvest

In more recent times, the so-called "bush bean" has been developed which does not require support and has all its pods develop simultaneously (as opposed to pole beans which develop gradually). This makes the bush bean more practical for commercial production.

History

Beans are one of the longest-cultivated plants. Broad beans, also called fava beans, in their wild state the size of a small fingernail, were gathered in Afghanistan and the Himalayan foothills. In a form improved from naturally occurring types, they were grown in Thailand since the early seventh millennium BCE, predating ceramics. They were deposited with the dead in ancient Egypt. Not until the second millennium BCE did cultivated, large-seeded broad beans appear in the Aegean, Iberia and transalpine Europe. In the *Iliad* (8th century BCE) is a passing mention of beans and chickpeas cast on the threshing floor.

The Beaneater (1580-90) by Annibale Carracci

Cooked beans on toast

Beans were an important source of protein throughout Old and New World history, and still are today.

The oldest-known domesticated beans in the Americas were found in Guitarrero Cave, an archaeological site in Peru, and dated to around the second millennium BCE.

Most of the kinds commonly eaten fresh or dried, those of the genus *Phaseolus*, come originally from the Americas, being first seen by a European when Christopher Columbus, during his exploration of what may have been the Bahamas, found them growing in fields. Five kinds of *Phaseolus* beans were domesticated by pre-Columbian peoples: common beans (*Phaseolus vulgaris*) grown from Chile to the northern part of what is now the United States, and lima and sieva beans (*Phaseolus lunatus*), as well as the less widely distributed teparies (*Phaseolus acutifolius*), scarlet runner beans (*Phaseolus coccineus*) and polyanthus beans (*Phaseolus polyanthus*) One especially famous use of beans by pre-Columbian people as far north as the Atlantic seaboard is the "Three Sisters" method of companion plant cultivation:

In the New World, many tribes would grow beans together with maize (corn), and squash. The corn would not be planted in rows as is done by European agriculture, but in a checkerboard/hex fashion across a field, in separate patches of one to six stalks each.

Beans would be planted around the base of the developing stalks, and would vine their way up as the stalks grew. All American beans at that time were vine plants, "bush beans" having been bred only more recently. The cornstalks would work as a trellis for the beans, and the beans would provide much-needed nitrogen for the corn.

Squash would be planted in the spaces between the patches of corn in the field. They would be provided slight shelter from the sun by the corn, would shade the soil and reduce evaporation, and would deter many animals from attacking the corn and beans because their coarse, hairy vines and broad, stiff leaves are difficult or uncomfortable for animals such as deer and raccoons to walk through, crows to land on, etc.

Dry beans come from both Old World varieties of broad beans (fava beans) and New World varieties (kidney, black, cranberry, pinto, navy/haricot).

Beans are a heliotropic plant, meaning that the leaves tilt throughout the day to face the sun. At night, they go into a folded "sleep" position.

Types

Currently, the world genebanks hold about 40,000 bean varieties, although only a fraction are mass-produced for regular consumption.

Beans, average, canned, sugarfree	
Nutritional value per 100 g (3.5 oz)	
Energy	334 kJ (80 kcal)
Carbohydrates	10.5 g
Fat	0.5 g
Protein	9.6 g
• Units • μg = micrograms • mg = milligrams • IU = International units	
Percentages are roughly approximated using US recommendations for adults.	

Some bean types include:

- Vicia
    - *Vicia faba* (broad bean or fava bean)

Vicia faba or broad beans, known in the US as fava beans

- Phaseolus
  - *Phaseolus acutifolius* (tepary bean)
  - *Phaseolus coccineus* (runner bean)
  - *Phaseolus lunatus* (lima bean)
  - *Phaseolus vulgaris* (common bean; includes the pinto bean, kidney bean, black bean, Appaloosa bean as well as green beans, and many others)
  - *Phaseolus polyanthus* (a.k.a. *P. dumosus*, recognized as a separate species in 1995)
- Vigna
  - *Vigna aconitifolia* (moth bean)
  - *Vigna angularis* (adzuki bean)
  - *Vigna mungo* (urad bean)
  - *Vigna radiata* (mung bean)
  - *Vigna subterranea* (Bambara bean or ground-bean)
  - *Vigna umbellata* (ricebean)
  - *Vigna unguiculata* (cowpea; also includes the black-eyed pea, yardlong bean and others)
- Cicer
  - *Cicer arietinum* (chickpea or garbanzo bean)
- Pisum
  - *Pisum sativum* (pea)
- Lathyrus
  - *Lathyrus sativus* (Indian pea)
  - *Lathyrus tuberosus* (tuberous pea)
- Lens

Lentils

- *Lens culinaris* (lentil)

■ Lablab

- *Lablab purpureus* (hyacinth bean)

Hyacinth beans

■ Glycine

- *Glycine max* (soybean)

■ Psophocarpus

- *Psophocarpus tetragonolobus* (winged bean)

Psophocarpus tetragonolobus (winged bean)

- Cajanus
    - *Cajanus cajan* (pigeon pea)
- Mucuna
    - *Mucuna pruriens* (velvet bean)
- Cyamopsis
    - *Cyamopsis tetragonoloba* or (guar)
- Canavalia
    - *Canavalia ensiformis* (jack bean)
    - *Canavalia gladiata* (sword bean)
- Macrotyloma
    - *Macrotyloma uniflorum* (horse gram)
- *Lupinus* (lupin)
    - *Lupinus mutabilis* (tarwi)
    - *Lupinus albus* (lupini bean)
- Arachis
    - *Arachis hypogaea* (peanut)

Health Concerns

Toxins

Some kinds of raw beans contain a harmful tasteless toxin, lectin phytohaemagglutinin, that must be removed by cooking. Red and kidney beans are particularly toxic, but other types also pose risks of food poisoning. A recommended method is to boil the beans for at least ten minutes; undercooked beans may be more toxic than raw beans.

Cooking beans, without bringing them to the boil, in a slow cooker at a temperature well below boiling may not destroy toxins. A case of poisoning by butter beans used to make falafel was reported; the beans were used instead of traditional broad beans or chickpeas, soaked and ground without boiling, made into patties, and shallow fried.

Bean poisoning is not well known in the medical community, and many cases may be misdiagnosed or never reported; figures appear not to be available. In the case of the UK National Poisons Information Service, available only to health professionals, the dangers of beans other than red beans were not flagged as of 2008.

Fermentation is used in some parts of Africa to improve the nutritional value of beans by removing toxins. Inexpensive fermentation improves the nutritional impact of flour from dry beans and improves digestibility, according to research co-authored by Emire Shimelis, from the Food Engineering Program at Addis Ababa University. Beans are a major source of dietary protein in Kenya, Malawi, Tanzania, Uganda and Zambia.

Bacterial Infection from Bean Sprouts

It is common to make beansprouts by letting some types of bean, often mung beans, germinate in moist and warm conditions; beansprouts may be used as ingredients in cooked dishes, or eaten raw or lightly cooked. There have been many outbreaks of disease from bacterial contamination, often by *salmonella*, *listeria*, and *Escherichia coli*, of beansprouts not thoroughly cooked, some causing significant mortality.

Antinutrients

Many types of bean contain significant amounts of antinutrients that inhibit some enzyme processes in the body. Phytic acid and phytates, present in grains, nuts, seeds and beans, interfere with bone growth and interrupt vitamin D metabolism. Pioneering work on the effect of phytic acid was done by Edward Mellanby from 1939.

Nutrition

Beans have significant amounts of fiber and soluble fiber, with one cup of cooked beans providing between nine and 13 grams of fiber. Soluble fiber can help lower blood cholesterol. Beans are also high in protein, complex carbohydrates, folate, and iron.

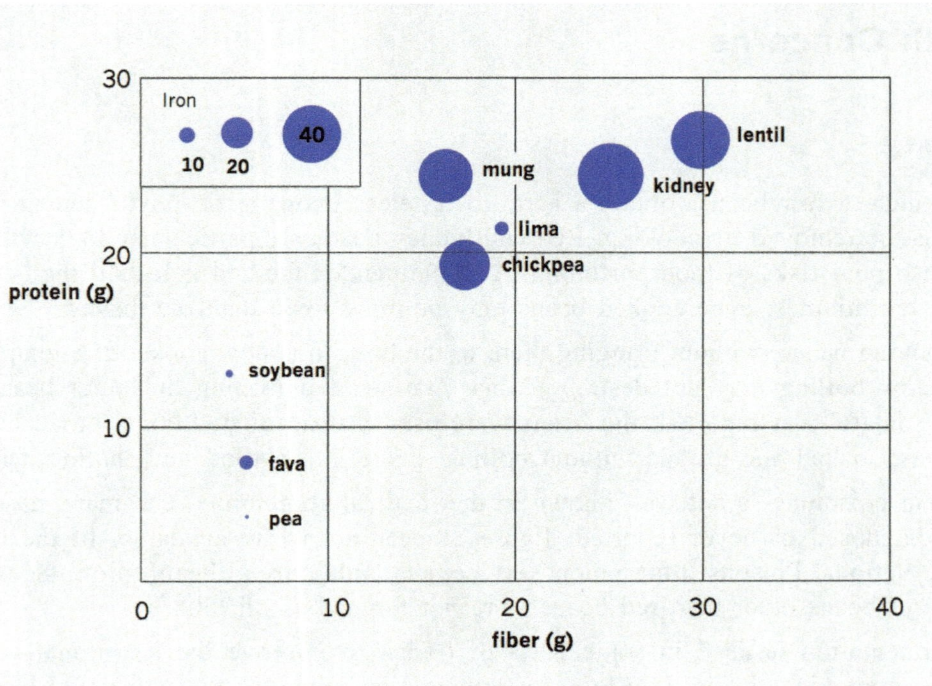

This figure shows the grams of fiber and protein per 100 gram serving of each legume. The size of the circle is proportional to its iron content. From this view, lentil and kidney beans are by far the healthiest while soybeans and peas have the least nutrients per serving.

Flatulence

Many edible beans, including broad beans and soybeans, contain oligosaccharides (particularly raffinose and stachyose), a type of sugar molecule also found in cabbage. An anti-oligosaccharide enzyme is necessary to properly digest these sugar molecules. As a normal human digestive tract does not contain any anti-oligosaccharide enzymes, consumed oligosaccharides are typically digested by bacteria in the large intestine. This digestion process produces flatulence-causing gases as a byproduct. Since sugar dissolves in water, another method of reducing flatulence associated with eating beans is to drain the water in which the beans have been cooked.

Some species of mold produce alpha-galactosidase, an anti-oligosaccharide enzyme, which humans can take to facilitate digestion of oligosaccharides in the small intestine. This enzyme, currently sold in the United States under the brand-names Beano and Gas-X Prevention, can be added to food or consumed separately. In many cuisines beans are cooked along with natural carminatives such as anise seeds, coriander seeds and cumin.

One effective strategy is to soak beans in alkaline (baking soda) water overnight before rinsing thoroughly. Sometimes vinegar is added, but only after the beans are cooked as vinegar interferes with the beans' softening.

Fermented beans will usually not produce most of the intestinal problems that unfermented beans will, since yeast can consume the offending sugars.

Production

Lablab bean and bean flower cultivated in West Bengal, India

The world leader in production of dry beans is Burma, followed by India and Brazil. In Africa, the most important producer is Tanzania.

Top ten dry bean producers—2013				
Country	**Production (tonnes)**	**Footnote**		
Myanmar	3,800,000	F		
India	3,630,000			
Brazil	2,936,444	A		
People's Republic of China	1,400,000	*		
Mexico	1,294,634			
Tanzania	1,150,000	F		
United States	1,110,668			
Kenya	529,265	F		
Uganda	461,000	*		
Rwanda	438,236			
World	**23,139,004**	**A**		
No symbol = official figure, P = official figure, F = FAO estimate, * = Unofficial/Semi-official/ mirror data, C = Calculated figure A = Aggregate (may include official, semi-official or estimates);				

LENTIL

The lentil (*Lens culinaris*) is an edible pulse. It is a bushy annual plant of the legume family, known for its lens-shaped seeds. It is about 40 cm (16 in) tall, and the seeds grow in pods, usually with two seeds in each.

In South Asian cuisine, split lentils (often with their hulls removed) are known as lentils. They usually eaten with rice or rotta, is a dietary staple throughout regions of India, Pakistan, Bangladesh and Nepal. As a food crop, the majority of world production comes from Canada, India and Australia.

Background

Lentils have been part of the human diet since aceramic Neolithic times, being one of the first crops domesticated in the Near East. Archeological evidence shows they were eaten 9,500 to 13,000 years ago.

Lentil plants in the field before flowering

Lentil colors range from yellow to red-orange to green, brown and black. Lentils also vary in size, and are sold in many forms, with or without the skins, whole or split.

Types

Illustration of the lentil plant, 1885

Red and brown comparison

- Beluga, black, bead-like shape, named for resemblance to Beluga caviar
- Brown/Spanish pardina
- French green
- Puy lentils , *Lens esculenta puyensis*, (small dark speckled blue-green), Protected Designation of Origin name
- Dark/Light Green
- Indianhead
- Yellow/tan lentils (red inside)
- Red Chief (decorticated yellow lentils)
- Eston Green (Small green)
- Richlea (medium green)
- Laird (large green)
- Masoor (brown-skinned lentils which are orange inside)
- Petite crimson/red (decorticated masoor lentils)
- *Macachiados* (big Mexican yellow lentils)

Nutritional Value

According to the USDA National Nutrient Database, 100 g of raw lentils (variety unspecified) provide 353 calories; the same weight of cooked lentils provides 116 kcal. Raw lentils are 8% water, 63% carbohydrates including 11% dietary fiber, 25% protein and 1% fat (table). Lentils are a rich source (20% or more of the Daily Value, DV) of numerous essential nutrients, including folate (120% DV), thiamin (76% DV), pantothenic acid (43% DV), vitamin B6 (42% DV), phosphorus (40% DV), iron (50% DV) and zinc

(35%), among others (table). When lentils are cooked by boiling, protein content declines to 9% of total composition, and B vitamins and minerals decrease due to the overall water content increasing (protein itself isn't lost).

Lentils, raw (dry weight)	
Nutritional value per 100 g (3.5 oz)	
Energy	1,477 kJ (353 kcal)
Carbohydrates	63 g
Sugars	2 g
Dietary fiber	10.7 g
Fat	1 g
Protein	25 g
Vitamins	
Thiamine (B1)	(76%) 0.87 mg
Riboflavin (B2)	(18%) 0.211 mg
Niacin (B3)	(17%) 2.605 mg
Pantothenic acid (B5)	(43%) 2.14 mg
Vitamin B6	(42%) 0.54 mg
Folate (B9)	(120%) 479 μg
Vitamin C	(5%) 4.5 mg
Minerals	
Calcium	(6%) 56 mg
Iron	(50%) 6.5 mg

Magnesium	(13%) 47 mg
Phosphorus	(40%) 281 mg
Potassium	(14%) 677 mg
Sodium	(0%) 6 mg
Zinc	(35%) 3.3 mg
Other constituents	
Water	8.3 g
Link to Full USDA Database entry	
• Units • µg = micrograms • mg = milligrams • IU = International units	
Percentages are roughly approximated using US recommendations for adults.	

Lentils have the second-highest ratio of protein per calorie of any legume, after soybeans.

The low levels of readily digestible starch (5%), and high levels of slowly digested starch, make lentils of potential value to people with diabetes. The remaining 65% of the starch is a resistant starch classified as RS1. A minimum of 10% in starch from lentils escapes digestion and absorption in the small intestine (therefore called "resistant starch").

Lentils also have anti-nutrient factors, such as trypsin inhibitors and a relatively high phytate content. Trypsin is an enzyme involved in digestion, and phytates reduce the bioavailability of dietary minerals. The phytates can be reduced by prolonged soaking and fermentation or sprouting.

Production

Lentils are relatively tolerant to drought, and are grown throughout the world. FAOSTAT reported that the world production of lentils for calendar year 2013 was 4,975,621 metric tons, primarily coming from Canada, India and Australia.

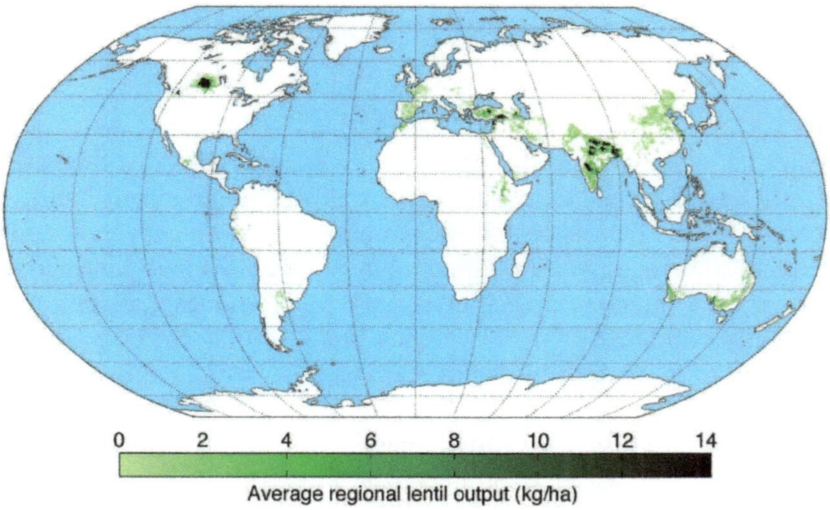

Worldwide lentil production

About a quarter of the worldwide production of lentils is from India, most of which is consumed in the domestic market. Canada is the largest export producer of lentils in the world, and Saskatchewan is the most important producing region in Canada (growing 99% of Canadian lentils). Statistics Canada estimates that Canadian lentil production for the 2009/10 year was a record 1.5 million metric tons. The most commonly grown type is the Laird lentil.

The Palouse region of eastern Washington and the Idaho panhandle, with its commercial center at Pullman, Washington, constitute the most important lentil-producing region in the United States.Montana and North Dakota are also significant lentil growers. The National Agricultural Statistics Service reported United States 2007 production at 154.5 thousand metric tons.

Top Lentil Producing Countries, 2014	
Country	**Metric Tons**
Canada	1,987,000
India	1,100,000
Australia	348,080
Turkey	345,000
Nepal	226,830
World	4,885,271
Source: UN Food & Agriculture Organization, Statistics Division	

Diseases

As food

The seeds require a cooking time of 10 to 40 minutes, depending on the variety—shorter for small varieties with the husk removed, such as the common red lentil—and have a distinctive, earthy flavor. Lentils with husk remain whole with moderate cooking; lentils without husk tend to disintegrate into a thick purée, which leads to quite different dishes. Lentil recipes are used throughout South Asia, the Mediterranean regions and West Asia. They are frequently combined with rice, which has a similar cooking time. A lentil and rice dish is referred to in Arab countries as *mujaddara* or *mejadra*. In Iran, rice and lentil is served with fried raisin; this dish is called Adas Polo, and usually comes with ground beef. Rice and lentils are also cooked together in *khichdi*, a popular dish in the Indian subcontinent (India and Pakistan); a similar dish, *kushari*, made in Egypt, is considered one of two national dishes. Lentils are used to prepare an inexpensive and nutritious soup all over Europe and North and South America, sometimes combined with some form of chicken or pork.

Dried lentils can be sprouted by soaking in water for one day and keeping moist for several days, which changes their nutrition profile and reduces undesirable phytic acid.

Lentils were a chief part of the diet of ancient Iranians, who consumed lentils daily in the form of a stew poured over rice.

Lentils are commonly eaten in Ethiopia in a stew-like dish called *kik*, or *kik wot*, one of the dishes people eat with Ethiopia's national food, *injera* flat bread. Yellow lentils are used to make a non-spicy stew, which is one of the first solid foods Ethiopian women feed their babies.

In the Indian Subcontinent lentil curry is part of the everyday diet, eaten with both rice and roti. Boiled lentils and lentil stock are used to thicken most vegetarian curries. They are also used as stuffing in dal parathas and Poori. Lentils are also used in many regional varieties of sweets.

In culture

The word "lens" in optics comes from the lentil-like shape of early double-convex lenses; lentil is *lens* in Latin, and gives its name to the botanical genus *Lens*.

In Italy and Hungary, eating lentils on New Year's Eve traditionally symbolizes the hope for a prosperous new year, most likely because of their round, coin-like form.

In Jewish mourning tradition, lentils and boiled eggs are traditional as food for mourners, because their round shape symbolizes the life cycle from birth to death.

REFERENCES

1. Wojciechowski, M. F.; Lavin, M.; Sanderson, M. J. (2004). "A phylogeny of legumes (Leguminosae) based on analysis of the plastid matK gene resolves

many well-supported sub clades within the family". American Journal of Botany. 91 (11): 1846–62. doi:10.3732/ajb.91.11.1846. PMID 21652332.

2. Angiosperm Phylogeny Group (2009). "An update of the Angiosperm Phylogeny Group classification for the orders and families of flowering plants: APG III" (PDF). Botanical Journal of the Linnean Society. 161 (2): 105–121. doi:10.1111/j.1095-8339.2009.00996.x. Retrieved 4 February 2014.
3. Watson L.; Dallwitz, M. J. (2007-06-01). "The families of flowering plants: Leguminosae". Retrieved 9 February 2008.
4. "GRIN-CA". Retrieved 1 September 2002.
5. Schrire, B. D.; Lewis, G. P.; Lavin, M. (2005). "Biogeography of the Leguminosae". In Lewis, G; Schrire, G.; Mackinder, B.; Lock, M. Legumes of the world. Kew, England: Royal Botanic Gardens. pp. 21–54. ISBN 1-900347-80-6.
6. International Code of Nomenclature for algae, fungi, and plants. Article 18.5 states: "The following names, of long usage, are treated as validly published: ....Leguminosae (nom. alt.: Fabaceae; type: Faba Mill. [= Vicia L.]); ... When the Papilionaceae are regarded as a family distinct from the remainder of the Leguminosae, the name Papilionaceae is conserved against Leguminosae."
7. Christenhusz, M. J. M.; Byng, J. W. (2016). "The number of known plants species in the world and its annual increase". Phytotaxa. Magnolia Press. 261 (3): 201–217. doi:10.11646/phytotaxa.261.3.1.
8. Judd, W. S., Campbell, C. S. Kellogg, E. A. Stevens, P.F. Donoghue, M. J. (2002), Plant systematics: a phylogenetic approach, Sinauer Axxoc, 287-292. ISBN 0-87893-403-0.
9. Stevens, P. F. "Fabaceae". Angiosperm Phylogeny Website. Version 7 May 2006. Retrieved 28 April 2008.
10. Magallón, S. A., and Sanderson, M. J.; Sanderson (2001). "Absolute diversification rates in angiosperm clades" (PDF). Evolution. 55 (9): 1762–1780. doi:10.1111/j.0014-3820.2001.tb00826.x. PMID 11681732.
11. Burnham, R. J., & Johnson, K. R. 2004. South American palaeobotany and the origins of neotropical rain forests. Phil. Trans. Roy. Soc. London B, 359: 1595–1610.
12. Lewis G., Schrire B., Mackinder B. and Lock M. 2005. (eds.) Legumes of the world. The Royal Botanic Gardens, Kew, Reino Unido. 577 pages. 2005. ISBN 1-900347-80-6.
13. Doyle, J. J., J. A. Chappill, C.D. Bailey, & T. Kajita. 2000. Towards a comprehensive phylogeny of legumes: evidence from rbcL sequences and non-molecular data. pp. 1 -20 in Advances in legume systematics, part 9, (P. S. Herendeen and A. Bruneau, eds.). Royal Botanic Gardens, Kew, UK.
14. Kajita, T.; Ohashi, H.; Tateishi, Y.; Bailey, C. D.; Doyle, J. J. (2001). "rbcL and legume phylogeny, with particular reference to Phaseoleae, Millettieae, and allies". Systematic Botany. 26: 515–536.

15. Wojciechowski, M. F., M. Lavin and M. J. Sanderson; Lavin; Sanderson (2004). "A phylogeny of legumes (Leguminosae) based on analysis of the plastid matK gene resolves many well-supported sub clades within the family". American Journal of Botany. 91 (11): 1846–1862. doi:10.3732/ajb.91.11.1846. PMID 21652332.

16. Angiosperm Phylogeny Group [APG] (2003). "An update of the Angiosperm Phylogeny Group classification for the orders and families of flowering plants: APG II" (PDF). Botanical Journal of the Linnean Society. 141: 399–436. doi:10.1046/j.1095-8339.2003.t01-1-00158.x.

17. Burkart, A. Leguminosas. In: Dimitri, M. 1987. Enciclopedia Argentina de Agricultura y Jardinería. Tomo I. Descripción de plantas cultivadas. Editorial ACME S.A.C.I., Buenos Aires. pages: 467-538.

18. Hélène L. Citerne; R. Toby Pennington; Quentin C. B. Cronk (8 August 2006). "An apparent reversal in floral symmetry in the legume Cadia is a homeotic transformation". PNAS. 103 (32): 12017–12020. doi:10.1073/pnas.0600986103. PMC 1567690Freely accessible. PMID 16880394.

19. Lloret, L. & Martínez-Romero, E. 2005. Evolución y filogenia de Rhizobium. Vol. 47, No. 1-2 pp. 43 – 60.

20. Sprent, J. I. 2001. Nodulation in legumes. Royal Botanic Gardens, Kew, UK.

21. Herendeen, P. S., W. L. Crepet, and D. L. Dilcher. 1992. The fossil history of the Leguminosae: phylogenetic and biogeographic implications. Pages 303 – 316 in Advances in Legume Systematics, part 4, the fossil record (P. S. Herendeen and D .L. Dilcher, eds). Royal Botanic Gardens, Kew, UK.

22. Crepet, W. L.; Taylor, D. W. (1985). "The Diversification of the Leguminosae: First Fossil Evidence of the Mimosoideae and Papilionoideae". Science. 228 (4703): 1087–1089. doi:10.1126/science.228.4703.1087. ISSN 0036-8075.

23. Crepet, W. L., and D. W. Taylor (1986). "Primitive mimosoid flowers from the Palaeocene-Eocene and their systematic and evolutionary implications.". American J. Botany. 73: 548–563. doi:10.2307/2444261. JSTOR 2444261.

24. Crepet, W. L., and P. S. Herendeen. 1992. Papilionoid flowers from the early Eocene of south eastern North America. Pages 43–55 in Advances in Legume Systematics, part 4, the fossil record (P. S. Herendeen and D. L. Dilcher, eds.). Royal Botanic Gardens, Kew, UK.

25. Herendeen, P. S. 1992. The fossil history of Leguminosae from the Eocene of south eastern North America. Pages 85-160 in Advances in Legume Systematics, part 4, the fossil record (Herendeen, P. S., and D. L. Dilcher, eds.). Royal Botanic Gardens, Kew, UK.

26. Herendeen, P. S. 2001. The fossil record of the Leguminosae: recent advances. In Legumes Down Under: the Fourth International Legume conference, Abstracts, 34–35. Australian National University, Canberra, Australia.

27. Herendeen, P. S., and S. Wing. 2001. Papilionoid legume fruits and leaves from the Palaeocene of north western Wyoming. Botany 2001 Abstracts, published by Botanical Society of America (http://www.botany2001.org/).

28. Wing, S. L., F. Herrera, and C. Jaramillo. 2004. A Palaeocene flora from the Cerrajón Formation, Guajíra Peninsula, north eastern Colombia. Pages 146-147 in VII International Organization of Paleobotany Conference Abstracts (21–26 March). Museo Egidio Feruglio, Trelew, Argentina.

29. Bruneau, A., Lewis, G. P., Herendeen, P. S., Schrire, B., & Mercure, M. 2008b. Biogeographic patterns in early-diverging clades of the Leguminosae. Pp. 98-99, in Botany 2008. Botany without Borders. [Botanical Society of America, Abstracts.]

30. Bruneau, A.; Mercure, M.; Lewis, G. P. & Herendeen, P. S. (2008). "Phylogenetic patterns and diversification in the caesalpinioid legumes". Canadian Journal of Botany. 86 (7): 697–718. doi:10.1139/B08-058.

31. Lavin, M., Herendeen, P. S., y Wojciechowski, M. F.; Herendeen; Wojciechowski (2005). "Evolutionary Rates Analysis of Leguminosae Implicates a Rapid Diversification of Lineages during the Tertiary". Systematic Biology. 54 (4): 575–594. doi:10.1080/10635150590947131. PMID 16085576.

32. Wikström, N. et ál. (2001), Savolainen, V., & Chase, M. W. 2001. Evolution of the angiosperms: Calibrating the family tree. Proc. Roy. Soc. London B, 268: 2211–2220.

33. Wojciechowski, M. F. 2003. Reconstructing the phylogeny of legumes (Leguminosae): An early 21st century perspective. Pp. 5-35, in Klitgaard, B. B. & Bruneau, A. (eds), Advances in Legume Systematics, Part 10, Higher Level Systematics. Royal Botanic Gardens, Kew.

34. Wojciechowski, M. F. 2004. Astragalus (Fabaceae): A molecular phylogenetic perspective. Brittonia 57: 382-396.

35. Wojciechowski, M. F. Sanderson, M. J., Baldwin, B. G., & Donoghue, M. J. 1993. Monophyly of aneuploid Astragalus: Evidence from nuclear ribosomal DNA internal transcribed spacer sequences. American J. Bot. 80: 711-722.

36. Wojciechowski, Martin F., Johanna Mahn, and Bruce Jones. 2006. Fabaceae. legumes. Version 14 June 2006. The Tree of Life Web Project, http://tolweb.org/

37. Schrire, B. D.; Lavin, M.; Lewis, G. P. (2005). Global distribution patterns of the Leguminosae: insights from recent phylogenies. In Friis, I; Balslev, H. Plant diversity and complexity patterns: local, regional and global dimensions. Biologiske Skrifter. 55. Viborg, Denmark: Special-Trykkeriet Viborg A/S. pp. 375–422. ISBN 87-7304-304-4.

38. Pan, Aaron D.; Jacobs, Bonnie F.; Herendeen, Patrick S. (2010). Detarieae sensu lato (Fabaceae) from the Late Oligocene (27.23 Ma) Guang River flora of north-western Ethiopia. Botanical Journal of the Linnean Society. 163: 44–54. doi:10.1111/j.1095-8339.2010.01044.x.

39. Doyle, J. J.; Luckow, MA (2003). The Rest of the Iceberg. Legume Diversity and Evolution in a Phylogenetic Context. Plant Physiology. 131 (3): 900–10. doi:10.1104/pp.102.018150. PMC 1540290Freely accessible. PMID 12644643.

40. Yokota, Keisuke; Hayashi, Makoto (2011). Function and evolution of nodulation genes in legumes. Cellular and Molecular Life Sciences. 68 (8): 1341–51. doi:10.1007/s00018-011-0651-4. PMID 21380559.

41. Markmann, Katharina; Giczey, Gábor; Parniske, Martin (2008). Functional Adaptation of a Plant Receptor- Kinase Paved the Way for the Evolution of Intracellular Root Symbioses with Bacteria. PLoS Biology. 6 (3): e68. doi:10.1371/journal.pbio.0060068. PMC 2270324Freely accessible. PMID 18318603.

42. Rodríguez-Llorente, Ignacio D.; Pérez-Hormaeche, Javier; Mounadi, Kaoutar El; Dary, Mohammed; Caviedes, Miguel A.; Cosson, Viviane; Kondorosi, Adam; Ratet, Pascal; Palomares, Antonio J. (2004). From pollen tubes to infection threads: Recruitment of Medicago floral pectic genes for symbiosis. The Plant Journal. 39 (4): 587–98. doi:10.1111/j.1365-313X.2004.02155.x. PMID 15272876.

43. Downie, J. Allan (2005). Legume Haemoglobins: Symbiotic Nitrogen Fixation Needs Bloody Nodules. Current Biology. 15 (6): R196–8. doi:10.1016/j.cub.2005.03.007. PMID 15797009.

44. Käss E, Wink M (1996). Molecular evolution of the Leguminosae: phylogeny of the three subfamilies based on rbcL sequences. Biochemical Systematics and Evolution. 24 (5): 365–378. doi:10.1016/0305-1978(96)00032-4.

45. Käss E, Wink M (1997). Phylogenetic relationships in the Papilionoideae (Family Leguminosae) based on nucleotide sequences of cpDNA (rbcL) and ncDNA (ITS1 and 2). Mol. Phylogenet. Evol. 8 (1): 65–88. doi:10.1006/mpev.1997.0410. PMID 9242596.

46. Doyle JJ, Doyle JL, Ballenger JA, Dickson EE, Kajita T, Ohashi H (1997). A phylogeny of the chloroplast gene rbcL in the Leguminosae: taxonomic correlations and insights into the evolution of nodulation. Am. J. Bot. 84 (4): 541–554. doi:10.2307/2446030. PMID 21708606.

47. Lavin M, Doyle JJ, Palmer JD (1990). Evolutionary significance of the loss of the chloroplast-DNA inverted repeat in the Leguminosae subfamily Papilionoideae. Evolution. 44 (2): 390–402. doi:10.2307/2409416. JSTOR 2409416.

48. Sanderson MJ, Wojciechowski MF (1996). Diversification rates in a temperate legume clade: are there so many species of Astragalus (Fabaceae)?. Am. J. Bot. 83 (11): 1488–1502. doi:10.2307/2446103. JSTOR 2446103.

49. Chappill JA. (1995). Cladistic analysis of the Leguminosae: the development of an explicit hypothesis. In Crisp MD, Doyle JJ. Advances in Legume Systematics, Part 7: Phylogeny. Royal Botanic Gardens, Kew, UK. pp. 1–10. ISBN 9780947643799.

50. Bruneau A, Mercure M, Lewis GP, Herendeen PS (2008). Phylogenetic patterns and diversification in the caesalpinioid legumes. Botany. 86 (7): 697–718. doi:10.1139/B08-058.

51. Cardoso D, Pennington RT, de Queiroz LP, Boatwright JS, Van Wykd BE, Wojciechowskie MF, Lavin M (2013). Reconstructing the deep-branching relationships of the papilionoid legumes. S. Afr. J. Bot. 89: 58–75. doi:10.1016/j.sajb.2013.05.001.

52. NOTE: The subfamilial name Papilionoideae for Faboideae is approved by the International Code of Nomenclature for algae, fungi, and plants, Article 19.8
53. Martin F. Wojciechowski; Johanna Mahn; Bruce Jones (2006). Fabaceae. The Tree of Life Web Project.
54. Allen, O. N., & E. K. Allen. 1981. The Leguminosae, A Source Book of Characteristics, Uses, and Nodulation. The University of Wisconsin Press, Madison, USA.
55. Duke, J. A. 1992. Handbook of Legumes of Economic Importance. Plenum Press, New York, USA.
56. Graham, P. H.; Vance, C. P. (2003). Legumes: importance and constraints to greater use. Plant Physiology. 131: 872–877. doi:10.1104/pp.017004.
57. Wojciechowski, M.F. 2006. Agriculturally & Economically Important Legumes.. Accessed 15 November 2008.
58. Sprent, Janet I. (2009). Legume Nodulation: A Global Perspective. Ames, Iowa: Wiley-Blackwell. p. 12. ISBN 1-4051-8175-3. Preview available at Google Books.
59. The gene bank and breeding of grain legumes (lupine, vetch, soya and beah) / B.S. Kurlovich and S.I. Repyev (Eds.), - St. Petersburg, The N.I. Vavilov Institute of Plant Industry, 1995, 438p. - (Theoretical basis of plant breeding. V.111)
60. Goulson, J.; Kaden, J.C.; Lepais, G.C. (2011). Population Structure, Dispersal and Colonization History of the Garden Bumblebee Bombus Hortorum in the Western Isles of Scotland. Conservation Genetics. 12: 867–879. doi:10.1007/s10592-011-0190-4.
61. Kuklinski, C. 2000. Farmacognosia : estudio de las drogas y sustancias medicamentosas de origen natural. Ediciones Omega, Barcelona. ISBN 84-282-1191-4
62. Marquez, A. C., Lara, O.F., Esquivel, R. B. & Mata, E. R. 1999. Composición, usos y actividad biológica: Plantas medicinales de México II. UNAM. First edition. México, D.F.
63. Macaya J. 1999. Leguminosas arbóreas y arbustivas cultivadas en Chile. Chloris Chilensis Año 2. N°1.
64. Ministerio de Educación de la Nación. Subsecretaría de Coordinación Administrativa. Día de la Flor Nacional El Ceibo. Efemérides Culturales Argentinas. Consulted 3 March 2010.
65. Gilbert Vargas Ulate. 1997. Geografía turística de Costa Rica. EUNED, 180 p. ISBN 9977-64-900-6, 9789977649009.
66. Lei N° 6.607, de 7 de dezembro de 1978. O Presidente da República, faço saber que o Congresso Nacional decreta e eu sanciono a seguinte Lei: Art. 1°- É declarada Árvore Nacional a leguminosa denominada Pau-Brasil (Caesalpinia echinata, Lam), cuja festa será comemorada, anualmente, quando o Ministério da Educação e Cultura promoverá campanha elucidativa sobre a relevância daquela espécie vegetal na História do Brasil.

67. Boden, Anne (1985). Golden Wattle: Floral Emblem of Australia (http). Australian National Botanic Gardens. Retrieved 8 October 2008.
68. Williams, Martin (1999). Golden Enigmatic Beauty (http). Bahuninia. Retrieved 8 October 2008.
69. The Plant List: A Working List of All Plant Species. Retrieved 7 March 2015.
70. Oxford English Dictionary - Pea
71. Rogers, Speed (2007). Man and the Biological World Read Books. pp. 169–170. ISBN 978-1-4067-3304-4 retrieved on 2009-04-15.
72. Pea. purdue.edu.
73. Zohary, Daniel and Hopf, Maria (2000). Domestication of Plants in the Old World, third edition. Oxford: University Press. ISBN 978-0-19-850356-9 p. 105–107
74. Pea Golden Podded, The Diggers Club
75. Purple podded peas
76. Dry Field Pea. purdue.edu.
77. Maguelonne Toussaint-Samat, A History of Food, 2nd ed. 2009:38ff.
78. Peas have been found in the Neolithic site of Abeurador in the south of France (Toussaint-Samat).
79. Theophrastus, Historia Plantarum, VIII.i.4.
80. Edict quoted in Michel Pitrat and Claude Four, Histoires de légumes: Des origines à l›orée du XXIe siècle, Le pois au cours des siècles :353.
81. Pitrat and Four
82. An account is in Toussaint-Samat.
83. Quoted by Michel Pitrat and Claude Four.
84. Bianchini, F.; Corbetta, F. (1976), The Complete Book of Fruits and Vegetables, New York: Crown, p. 40, ISBN 978-0-517-52033-8
85. Hedrick, U.P. (1919), Sturtevant›s Notes on Edible Plants, Report of the New York Agricultural Experiment Station for the Year 1919 II, Albany: J.B Lyon Company, State Printers, archived from the original on July 18, 2011, retrieved Feb 26, 2010
86. Kafka, B. (2005), Vegetable Love, New York: Artisan, p. 297, ISBN 978-1-57965-168-8
87. Product how-to guides. PCC Natural Markets.
88. Sanningen om ärtsoppan" (Swedish)
89. Wainwright, Martin (2005-05-23). Onions come top for British palates. The Guardian. London.
90. Vrabel, Jeff (2016-07-10). If You Like Almond Milk, You›ll Love...Pea Milk?. Retrieved 2016-07-13.
91. How frozen vegetable is made - production process, making, used, processing, product, industry, machine. madehow.com.

92. Sivasankar, B. (2002). Food Processing and Preservation. PHI Learning Pvt. Ltd. pp. 175-177. ISBN 8120320867
93. Nutrition Facts: Peas. Nutrition. vegonline.org. Retrieved February 28, 2015.
94. Jegtvig, Shereen (July 17, 2007). Peas. Nutrition. About.com. Retrieved 2011-01-28.
95. Pownall TL, Udenigwe CC, Aluko RE (2010). Amino acid composition and antioxidant properties of pea seed (Pisum sativum L.) enzymatic protein hydrolysate fractions. Journal of Agricultural and Food Chemistry. 58 (8): 4712–4718. doi:10.1021/jf904456r. PMID 20359226.
96. Peas-Western Oregon, Commercial Vegetable Production Guides. Oregon State University.
97. Watson, Terumi. UT Garden›s Plant of the Month - Tom Thumb Pea. University of Tennessee. Retrieved February 28, 2015.
98. Vegetable Cultivar Descriptions for North America.
99. Kelvedon Wonder is popular in the United Kingdom, but uncommon elsewhere.
100. Recruit Peas. Pure Line Seeds. Retrieved February 28, 2015.
101. Hagedorn, D. J. (1976). Handbook of pea diseases (PDF). University of Wisconsin - Extension.
102. Hagedorn, Donald J. (1974). Virus Diseases of Pea, Pisum sativum. St. Paul, Minnesota: American Phytopathological Society. p. 7.
103. Barkley, Shelley (2007-05-02). Pea Leaf Weevil. Agriculture and Rural Development website. Government of Alberta. Retrieved 2009-04-05.
104. Gregor Mendel: The Pea Plant Experiment. juliantrubin.com.
105. The Garden Pea Retrieved 14 April 2012.
106. Pitman, Sean D. The Father of Genetics May 2002. Retrieved 14 April 2012.
107. Slanc et al. (2009). Screening of Selected Food and Medicinal Plant Extracts for Pancreatic Lipase Inhibition. Phytotherapy Research, 23:874-877 http://onlinelibrary.wiley.com/doi/10.1002/ptr.2718/abstract;jsessionid=962613A3CDE6F5E498E4D6C4B1860DE5.f01t01
108. Sanchez-Monge, R.; G. Lopez-Torrejon; C. Y. Pascual; J. Varela; M. Martin-Esteban; G. Salcedo (2004). Vicilin and convicilin are potential major allergens from pea. Clinical & Experimental Allergy. 34 (11): 1747–1753. doi:10.1111/j.1365-2222.2004.02085.x. ISSN 0954-7894.
109. Glucose-6-Phosphate Dehydrogenase Deficiency. TheFreeDictionary.com.
110. favism - genetic disorder. Encyclopedia Britannica.
111. The Nitrogen cycle and Nitrogen fixation, Jim Deacon, Institute of Cell and Molecular Biology, The University of Edinburgh
112. Postgate, J (1998). Nitrogen Fixation, 3rd Edition. Cambridge University Press, Cambridge UK

113. Smil, V (2000). Cycles of Life. Scientific American Library.
114. Oxford English Dictionary
115. πίσος, Henry George Liddell, Robert Scott, A Greek-English Lexicon, on Perseus Digital Library

CHAPTER 2

Farming of Solanaceae Family

Chapter Contents

Solanaceae is a family of flowering plants in the Solanales order, characterized by five-petaled flowers, typically conical or funnelform, and alternate or alternate to opposite leaves, and including some of the more important food and drug-producing plants, including potato, tomato, eggplant, tobacco, chili pepper, and deadly nightshade. This family is known as the potato or nightshade family, and the name nightshade sometimes is used as the common name in general for plants of this family, although often the name nightshade is reserved for members of the genus Solarum. Solanaceae includes over 3,000 species of annual plants and perennial plants, vines, herbaceous plants, sub-shrubs, shrubs, and some trees.

The Solanaceae family is characteristically ethnobotanical, that is, extensively utilized by humans. It is an important source of food, spice, and medicine, and many plants are used as ornamentals, including the petunia and butterfly flower. In economic importance, Solanaceae is the third most important plant taxon and the most valuable in terms of vegetable crops (SGN). It also offers the most variability in terms of crop species, including tuber-bearing crops (potato), fruit-bearing vegetables (tomato, eggplant, peppers), leaf vegetables (Solanum aethiopicum, S. macrocarpon), and medicinal plants (eg. Capsicum) (SGN). Members of Solanaceae provide stimulants, poisons, narcotics, pain relievers, and so forth.

INTRODUCTION

The Solanaceae, or nightshades, are an economically important family of flowering plants. The family ranges from annual and perennial herbs to vines, lianas, epiphytes, shrubs, and trees, and includes a number of important agricultural crops, medicinal plants, spices, weeds, and ornamentals. Many members of the family contain potent alkaloids, and some are highly toxic, but many cultures eat nightshades, in some cases as staple foods. The family belongs to the order Solanales, in the asterid group and class Magnoliopsida (dicotyledons). The Solanaceae consists of about 98 genera and some 2,700 species,with a great diversity of habitats, morphology and ecology.

The name Solanaceae derives from the genus Solanum, "the nightshade plant". The etymology of the Latin word is unclear. The name may come from a perceived resemblance of certain solanaceous flowers to the sun and its rays. At least one species of Solanum is known as the "sunberry". Alternatively, the name could originate from the Latin verb solari, meaning "to soothe", presumably referring to the soothing pharmacological properties of some of the psychoactive species of the family.

A flowering Brugmansia suaveolens from the US Botanic Garden

The family has a worldwide distribution, being present on all continents except Antarctica. The greatest diversity in species is found in South America and Central America.

The Solanaceae include a number of commonly collected or cultivated species. The most economically important genus of the family is Solanum, which contains the potato (S. tuberosum, in fact, another common name of the family is the "potato family"), the tomato (S. lycopersicum), and the eggplant or aubergine (S. melongena). Another important genus, Capsicum, produces both chili peppers and bell peppers.

The genus Physalis produces the so-called groundcherries, as well as the tomatillo (Physalis philadelphica), the Cape gooseberry and the Chinese lantern. The genus Lycium contains the boxthorns and the wolfberry Lycium barbarum. Nicotiana contains, among other species, tobacco. Some other important members of Solanaceae include a number of ornamental plants such as Petunia, Browallia, and Lycianthes, and sources of psychoactive alkaloids, Datura, Mandragora (mandrake), and Atropa belladonna (deadly nightshade). Certain species are widely known for their medicinal uses, their psychotropic effects, or for being poisonous.

Fruits including tomatoes, tomatillos, eggplant, and chili peppers.

Most of the economically important genera are contained in the subfamily Solanoideae, with the exceptions of tobacco (Nicotiana tabacum, Nicotianoideae) and petunia (Petunia × hybrida, Petunioideae),

Many of the Solanaceae, such as tobacco and petunia, are used as model organisms in the investigation of fundamental biological questions at the cellular, molecular, and genetic levels.

ETYMOLOGY AND PRONUNCIATION

The name "Solanaceae" comes to international scientific vocabulary from New Latin, from Solanum, the type genus, + -aceae, a standardized suffix for plant family names in modern taxonomy. The genus name comes from the Classical Latin word solanum, referring to nightshades (especially Solanum nigrum), "probably from sol, 'sun', + -anum, neuter of -anus."

DESCRIPTION

Plants in the "Solanaceae" can take the form of herbs, shrubs, trees, vines and lianas, and sometimes epiphytes. They can be annuals, biennials, or perennials, upright or decumbent. Some have subterranean tubers. They do not have laticifers, nor latex, nor coloured saps. They can have a basal or terminal group of leaves or neither of these types. The leaves are generally alternate or alternate to opposed (that is, alternate at the base of the plant and opposed towards the inflorescence). The leaves can be herbaceous, leathery, or transformed into spines. The leaves are generally petiolate or subsessile, rarely sessile. They are frequently inodorous, but on occasions, they are aromatic or fetid. The foliar lamina can be either simple or compound, and the latter can be either pinnatifid or ternate. The leaves have reticulated venation and lack a basal meristem. The laminae are generally dorsiventral and lack secretory cavities. The stomata are generally confined to one of a leaf's two sides; they are rarely found on both sides.

Illustration of Solanum dulcamara, 1.- Flower, 2.- Flower in longitudinal section, without the petals; 3.- Androecium; 4.- Ovary, in transverse section; 5.- Seed viewed from above; 6.- Seed in transverse section, note the curved embryo surrounding the endosperm; A.- Branch with leaves and flowers; B.- Stem with immature and mature fruit

The flowers are generally hermaphrodites, although some are monoecious, andromonoecious, or dioecious species (such as some Solanum or Symonanthus). Pollination is entomophilous. The flowers can be solitary or grouped into terminal, cymose, or axillary inflorescences. The flowers are medium-sized, fragrant (Nicotiana), fetid (Anthocercis), or inodorous. The flowers are usually actinomorphic, slightly zygomorphic, or markedly zygomorphic (for example, in flowers with a bilabial corolla in Schizanthus species). The irregularities in symmetry can be due to the androecium, to the perianth, or both at the same time. In the great majority of species, the flowers have a differentiated perianth with a calyx and corolla (with five sepals and five petals, respectively) an androecium with five stamens and two carpels forming a gynoecium with a superior ovary (they are therefore referred to as pentamers and tetracyclic). The stamens are epipetalous and are typically present in multiples of four or five, most commonly four or eight. They usually have a hypogynous disk. The calyx is gamosepalous (as the sepals are joined together forming a tube), with the (4)5(6) segments equal, it has five lobes, with the lobes shorter than the tube, it is persistent and often accrescent. The corolla usually has five petals that are also joined together forming a tube. Flower shapes are typically rotate (wheel-shaped, spreading in one plane, with a short tube) or tubular (elongated cylindrical tube), campanulated or funnel-shaped.

The androecium has (2)(4)5(6) free stamens within it, oppositsepals (that is, they alternate with the petals), they are usually fertile or, in some cases (for example in Salpiglossideae) they have staminodes. In the latter case, there is usually either one staminode (Salpiglossis) or three (Schizanthus). The anthers touch on their upper end forming a ring, or they are completely free, dorsifixed, or basifixed with poricide dehiscence or through small longitudinal cracks. The stamen's filament can be filliform or flat. The stamens can be inserted inside the coralline tube or exserted. The plants demonstrate simultaneous microsporogenesis, the microspores are tetrad, tetrahedral, or isobilateral. The pollen grains are bicellular at the moment of dehiscence, usually open and angular.

The gynoecium is bicarpelar (rarely three- or five-locular) with a superior ovary and two locules, which may be secondarily divided by false septa, as is the case for Nicandreae and Datureae. The gynoecium is located in an oblique position relative to the flower's median plane. They have one style and one stigma; the latter is simple or bilobate. Each locule has one to 50 ovules that are anatropous or hemianatropous with axillar placentation. The development of the embryo sack can be the same as for Polygonum or Allium species. The embryo sack's nuclear poles become fused before fertilization. The three antipodes are usually ephemeral or persistent as in the case of Atropa. The fruit can be a berry as in the case of the tomato or wolfberry a dehiscent capsule as in Datura, or a drupe. The fruit has axial placentation. The capsules are normally septicidal or rarely loculicidal or valvate. The seeds are usually endospermic, oily (rarely starchy), and without obvious hairs. The seeds of most Solanaceae are round and flat, about 2–4 mm (0.079–0.157 in) in diameter. The embryo can be straight or curved, and has two cotyledons. Most species in the Solanaceae have 2n=24 chromosomes, but the number may be a higher multiple of 12 due to polyploidy. Wild potatoes, of which there are about 200, are predominantly diploid (2 × 12 = 24 chromosomes), but triploid (3 × 12 = 36 chromosomes), tetraploid (4 × 12 = 48 chromosomes), pentaploid (5 × 12 = 60) and even hexaploid (6 × 12 = 72 chromosome) species or populations exist. The cultivated species Solanum tuberosum has 4 × 12 = 48 chromosomes. Some Capsicum species have 2 × 12 = 24 chromosomes, while others have 26 chromosomes.

Diversity of Characteristics

Despite the previous description, the Solanaceae exhibit a large morphological variability, even in their reproductive characteristics. Examples of this diversity include:

- The number of carpels that form the gynoecium

In general, the Solanaceae have a gynoecium (the female part of the flower) formed of two carpels. However, Melananthus has a monocarpelar gynoecium, there are three or four carpels in Capsicum, three to five in Nicandra, some species of Jaborosa and Trianaea and four carpels in Iochroma umbellatum.

- The number of locules in the ovary

The number of locules in the ovary is usually the same as the number of carpels. However, some species occur in which the numbers are not the same due to the existence of false septa (internal walls that subdivide each locule), such as in Datura and some members of the Lycieae (the genera Grabowskia and Vassobia).

- Type of ovules and their number

The ovules are generally inverted, folded sharply backwards (anatropous), but some genera have ovules that are rotated at right angles to their stalk (campilotropous) as in Phrodus, Grabowskia or Vassobia), or are partially inverted (hemitropous as in Cestrum, Capsicum, Schizanthus and Lycium). The number of ovules per locule also varies from a few (two pairs in each locule in Grabowskia, one pair in each locule in Lycium) and very occasionally only one ovule is in each locule as for example in Melananthus.

- The type of fruit

The fruits of the great majority of the Solanaceae are berries or capsules (including pyxidia) and less often drupes. Berries are common in the subfamilies Cestroideae, Solanoideae (with the exception of Datura, Oryctus, Grabowskia and the tribe Hyoscyameae) and the tribe Juanulloideae (with the exception of Markea). Capsules are characteristic of the subfamilies Cestroideae (with the exception of Cestrum) and Schizanthoideae, the tribes Salpiglossoideae and Anthocercidoideae, and the genus Datura. The tribe Hyoscyameae has pyxidia. Drupes are typical of the Lycieae tribe and in Iochrominae.

Alkaloids

Alkaloids are nitrogenous organic substances produced by plants as a secondary metabolite and which have an intense physiological action on animals even at low doses. Solanaceae are known for having a diverse range of alkaloids. To humans, these alkaloids can be desirable, toxic, or both. The tropanes are the most well-known of the alkaloids found in the Solanaceae. The plants that contain these substances have been used for centuries as poisons. However, despite being recognized as poisons, many of these substances have invaluable pharmaceutical properties. Many species contain a variety of alkaloids that can be more or less active or poisonous, such as scopolamine, atropine, hyoscyamine, and nicotine. They are found in plants such as the henbane (Hyoscyamus albus), belladonna (Atropa belladonna), datura or jimson (Datura stramonium), mandrake (Mandragora autumnalis), tobacco, and others. Some of the main types of alkaloids are:

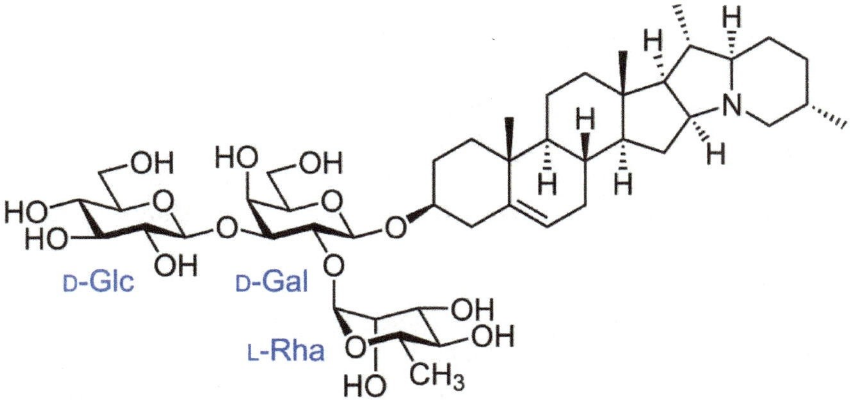

Chemical structure of solanine

- Solanine: A toxic glycoalkaloid with a bitter taste, it has the formula $C_{45}H_{73}NO_{15}$. It is formed by the alkaloid solanidine with a carbohydrate side chain. It is found in leaves, fruit, and tubers of various Solanaceae such as the potato and tomato. Its production is thought to be an adaptive defence strategy against herbivores. Substance intoxication from solanine is characterized by gastrointestinal disorders (diarrhoea, vomiting, abdominal pain) and neurological disorders (hallucinations and headache). The median lethal dose is between 2 and 5 mg per kg of body weight. Symptoms become manifest 8 to 12 hr after ingestion. The amount of these glycoalkaloids in potatoes, for example, varies significantly depending of environmental conditions during their cultivation, the length of storage, and the variety. The average glycoalkaloid concentration is 0.075 mg/g of potato. Solanine has occasionally been responsible for poisonings in people who ate berries from species such as Solanum nigrum or Solanum dulcamara, or green potatoes.

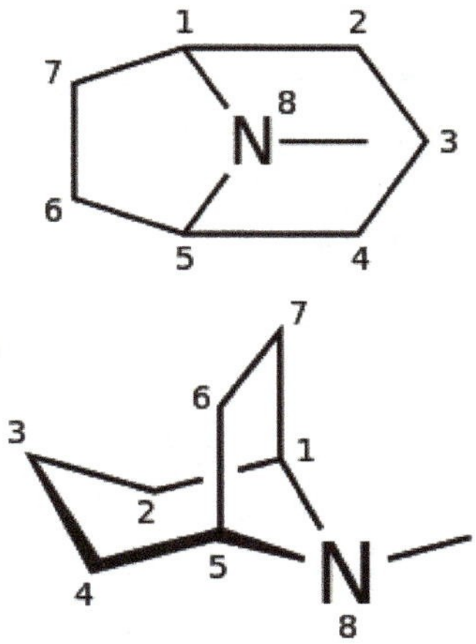

Chemical structure of the tropanes.

- Tropanes: The term "tropane" comes from a genus in which they are found, Atropa (the belladonna genus). Atropa is named after the Greek Fate, Atropos, who cut the thread of life. This nomenclature reflects its toxicity and lethality. They are bicyclic organic nitrogen compounds (IUPAC nomenclature: 8-Methyl-8-azabicyclo[3.2.1]octane), with the chemical formula of $C_8H_{15}N$. These alkaloids include, among others, atropine, cocaine, scopolamine, and hyoscyamine. They are found in various species, such as mandrake (Mandragora autumnalis), black henbane or stinking nightshade (Hyoscyamus niger), belladonna (Atropa belladonna) the stramonium (Datura stramonium) and Brugmansia species, as well as many others in the Solanaceae family. Pharmacologically, they are the most powerful known anticholinergics in existence, meaning they inhibit the neurological signals transmitted by the endogenous neurotransmitter, acetylcholine. More commonly, they can halt many types of allergic reactions. Symptoms of overdose may include dry mouth, dilated pupils, ataxia, urinary retention, hallucinations, convulsions, coma, and death. Atropine, a commonly used ophthalmological agent, dilates the pupils and thus facilitates examination of the interior of the eye. In fact, juice from the berries of A. belladonna were used by Italian courtesans during the Renaissance to exaggerate the size of their eyes by causing the dilation of their pupils. Despite the extreme toxicity of the tropanes, they are useful drugs when administered in extremely small dosages. They can reverse cholinergic poisoning, which can be caused by overexposure to organophosphate insecticides and chemical warfare agents such as sarin and VX. Scopolamine (found in Hyoscyamus muticus and Scopolia atropioides), is used as an antiemetic against motion sickness or for people suffering from nausea as a result of receiving chemotherapy. Scopolamine and hyoscyamine are the most widely used tropane alkaloids in pharmacology and medicine due to their effects on the parasympathetic nervous system. Atropine has a stimulant effect on the central nervous system and heart, whereas scopolamine has a sedative effect. These alkaloids cannot be substituted by any other class of compounds, so they are still in demand. This is one of the reasons for the development of an active field of research into the metabolism of the alkaloids, the enzymes involved, and the genes that produce them. Hyoscyamine 6-β hydroxylase, for example, catalyses the hydroxylation of hyoscyamine that leads to the production of scopolamine at the end of the tropane's biosynthetic pathway. This enzyme has been isolated and the corresponding gene cloned from three species: H. niger, A. belladonna and B. candida.

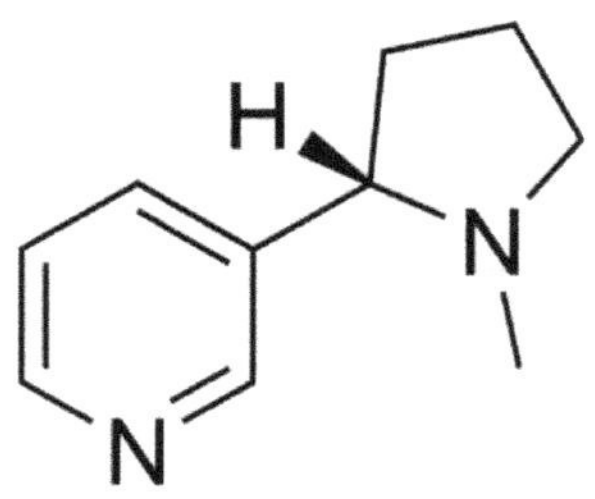

Chemical structure of nicotine.

- Nicotine: Nicotine (IUPAC nomenclature (S)-3-(1-methylpyrrolidin-2-il) pyridine) is a pyrrolidine alkaloid produced in large quantities in the tobacco plant (Nicotiana tabacum), but is also found in lower concentrations in other species such as the potato, tomato, and pepper. Its function in a plant is to act as a defence against herbivores, as it is an excellent neurotoxin, in particular against insects. In fact, nicotine has been used for many years as an insecticide, although its use is currently being replaced by synthetic molecules derived from its structure. At low concentrations, nicotine acts as a stimulant in mammals, which causes the dependency in smokers. Like the tropanes, it acts on cholinergic neurons, but with the opposite effect (it is an agonist as opposed to an antagonist). It has a higher specificity for nicotinic acetylcholine receptors than other ACh proteins.

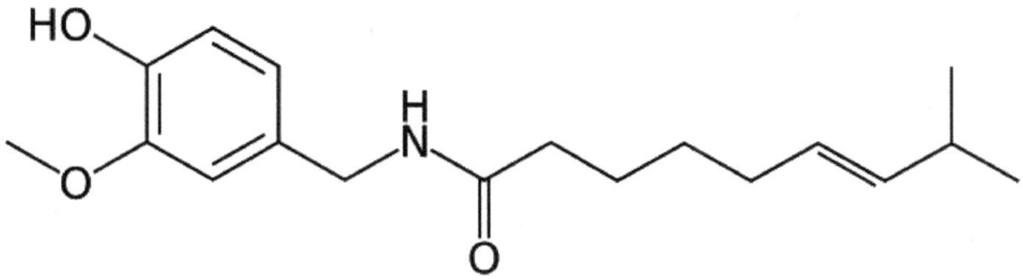

Chemical structure of capsaicin

- Capsaicin: Capsaicin (IUPAC nomenclature 8-methyl-N-vanillyl-trans-6-nonenamide) is structurally different from nicotine and the tropanes. It is found in species of the genus Capsicum, which includes chilies and habaneros and it is the active ingredient that determines the Scoville rating of these spices. The compound is not noticeably toxic to humans. However, it stimulates specific pain receptors in the majority of mammals, specifically those related to the perception of heat in the oral mucosa and other epithelial tissues. When capsaicin comes into contact with these mucosae, it causes a burning sensation little different from a burn caused by fire. Capsaicin affects only mammals, not birds. Pepper seeds can survive the digestive tracts of birds; their fruit becomes brightly coloured once its seeds are mature enough to germinate, thereby attracting the attention of birds that then distribute the seeds. Capsaicin extract is used to make pepper spray, a useful deterrent against aggressive mammals.

Distribution

Even though members of the Solanaceae are found on all continents except Antarctica, the greatest variety of species are found in Central America and South America. Centers of diversity also occur in Australia and Africa. Solanaceae occupy a great number of different ecosystems, from deserts to rainforests, and are often found in the secondary vegetation that colonizes disturbed areas. In general, plants in this family are of tropical and temperate distribution.

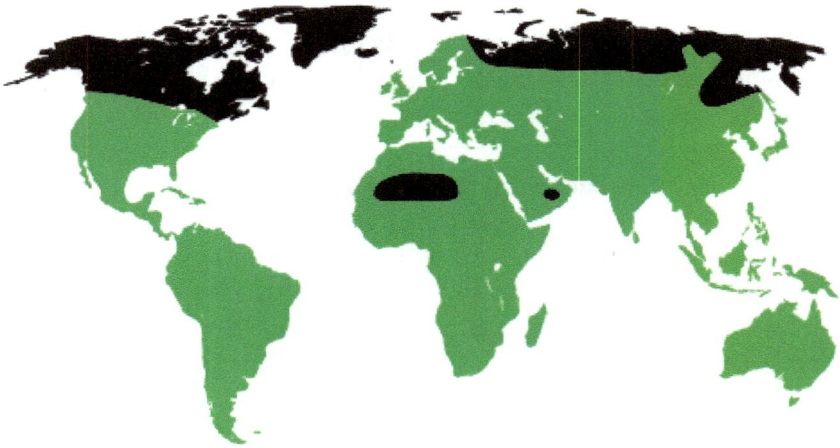

Map showing the distribution of the Solanaceae throughout the world (light green areas)

Taxonomy

The following taxonomic synopsis of the solanaceas, including subfamilies, tribes and genera, is based on the most recent molecular phylogenetics studies of the family:

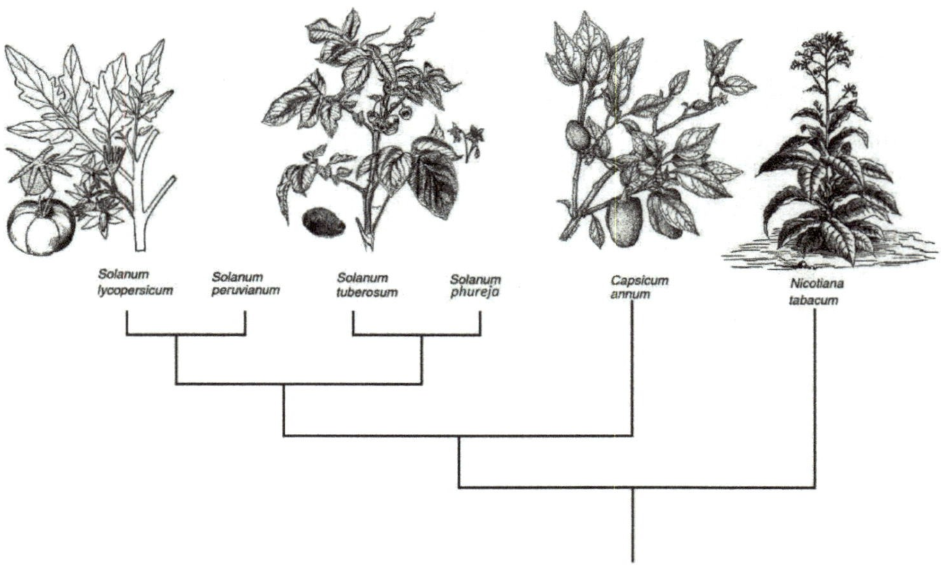

Cladogram showing the relationship between the three genera of the Solanaceae family

Cestroideae (Browallioideae)

This subfamily is characterised by the presence of pericyclic fibres, an androecium with four or five stamens, frequently didynamous. The basic chromosome numbers are highly variable, from x=7 to x=13. The subfamily consists of eight genera (divided into three tribes) and about 195 species distributed throughout the Americas. The Cestrum genus is the most important, as it contains 175 of the 195 species in the subfamily.

Cestrum elegans, a cestroidea used as an ornamental

Browallia americana

Flower of Salpiglossis sinuata, Botanischer Garten Jena, Germany

The Cestreae tribe is unusual because it includes taxa with long chromosomes (from 7.21 to 11.511 μm in length), when the rest of the family generally possesses short chromosomes (for example between 1.5 and 3.52 μm in the Nicotianoideae)

- Browallieae Hunz.
  - Browallia L., genus with six species distributed throughout the neotropic ecozone to Arizona in the United States
  - Streptosolen Miers, monotypic genus native to the Andes
- Cestreae tribe Don, three genera of woody plants, generally shrubs
  - Cestrum L., some 175 species distributed throughout the neotropic ecozone
  - Sessea Ruiz & Pav., 16 species from the Andes
  - Vestia Willd., monotypic genus from Chile
- Salpiglossideae tribe (Benth.) Hunz.
  - Reyesia Gay, four species, distributed throughout Argentina and Chile
  - Salpiglossis Ruiz & Pav., two species originating from southern South America

Goetzeoideae

This subfamily is characterized by the presence of drupes as fruit and seeds with curved embryos and large fleshy cotyledons. The basic chromosome number is x=13. It includes four genera and five species distributed throughout the Greater Antilles. Some authors suggest their molecular data indicate the monotypic genera Tsoala Bosser & D'Arcy should be included in this subfamily, endemic to Madagascar, and Metternichia to the southeast of Brazil. Goetzeaceae Airy Shaw is considered as a synonym of this subfamily.

- Coeloneurum Radlk., monotypic genus endemic to Hispaniola
- Espadaea Rchb., monotypic, from Cuba
- Goetzea Wydler, includes two species from the Antilles
- Henoonia Griseb., monotypic, originating in Cuba

Petunioideae

Molecular phylogenetics indicates that Petunioideae is the sister clade of the subfamilies with chromosome number x=12 (Solanoideae and Nicotianoideae). They contain calistegins, alkaloids similar to the tropanes. The androecium is formed of four stamens (rarely five), usually with two different lengths.

Nierenbergia frutescens, a petunoidea.

The basic chromosome number of this subfamily can be x=7, 8, 9 or 11. It consists of 13 genera and some 160 species distributed throughout Central and South America. Molecular data suggest the genera originated in Patagonia. Benthamiella, Combera, and Pantacantha form a clade that can be categorized as a tribe (Benthamielleae) that should be in the subfamily Goetzeoideae.

- Benthamiella Speg., 12 species native to Patagonia
- Bouchetia Dunal, three neotropical species
- Brunfelsia L., around 45 species from the neotropics
- Combera Sandw., two species from Patagonia
- Fabiana Ruiz & Pav., 15 species native to the Andes
- Hunzikeria D'Arcy, three species from the southwest United States and Mexico

- LatuaPhil., one species from the south of Chile
- Leptoglossis Benth., seven species from western South America
- Nierembergia Ruiz & Pav., 21 species from South America
- Pantacantha Speg., monospecific genus from Patagonia
- Calibrachoa Cerv. ex La Llave & Lex. consists of 32 species from the neotropics. The morphological data suggest this genus should be included within the Petunia. However, the molecular and cytogenetic data indicate both should be kept separate. In fact, Calibrachoa has a basic chromosome number x=9, while that of Petunia is x=7.
- Petunia (Juss.) Wijsman, 18 species from South America
- Plowmania Hunz. & Subils, monotypic genus from Mexico and Guatemala

Schizanthoideae

The Schizanthoideae include annual and biennial plants with tropane alkaloids, without pericyclic fibres, with characteristic hair and pollen grains. The flowers are zygomorphic. The androecium has two stamens and three stamenodes, anther dehiscence is explosive.

Zygomorphic flowers, with bilabiate corolla of Schizanthus pinnatus, a schizanthoidea ornamental

The embryo is curved. The basic chromosome number is x=10. Schizanthus is a somewhat atypical genus among the Solanaceae due to its strongly zygomorphic flowers and basic chromosome number. Morphological and molecular data suggest Schizanthus is a sister genus to the other Solanaceae and diverged early from the rest, probably in the late Cretaceous or in the early Cenozoic, 50 million years ago. The great diversity of flower types within Schizanthus has been the product of the species' adaptation to the different types of pollinators that existed in the Mediterranean, high alpine, and desert ecosystems then present in Chile and adjacent areas of Argentina.

- Schizanthus Ruiz & Pav., 12 species originating from Chile.

Schwenckioideae

Annual plants with pericyclic fibres, their flowers are zygomorphic, the androecium has four didynamous stamens or three stamenodes; the embryo is straight and short. The basic chromosome number is x=12. It includes four genera and some 30 species distributed throughout South America.

- Heteranthia Nees & Mart., one species from Brazil
- Melananthus Walp., five species from Brazil, Cuba, and Guatemala
- Protoschwenckia Soler , monotypic genus from Bolivia and Brazil, some molecular phylogenetic studies have suggested this genus has an uncertain taxonomic position within the subfamily
- Schwenckia L., 22 species distributed throughout the neotropical regions of America

Nicotianoideae

Tobacco inflorescence, Nicotiana tabacum

- Anthocercideae G. Don: This tribe, endemic to Australia, contains 31 species in seven genera. Molecular phylogenetic studies of the tribe indicate it is the sister of Nicotiana, and the genera Anthocercis, Anthotroche, Grammosolen, and Symonanthus are monophyletic. Some characteristics are also thought to be derived from within the tribe, such as the unilocular stamens with semicircular opercula, bracteolate flowers, and berries as fruit.
    - Anthocercis Labill., 10 species, Australia
    - Anthotroche Endl., four species, Australia
    - Crenidium Haegi, monotypic genus, Australia
    - Cyphanthera Miers, 9 species, Australia
    - Duboisia R.Br., four species, Australia
    - Gramnosolen Haegi, two species, Australia
    - Symonanthus Haegi, two species, Australia
- Nicotianeae tribe Dum.
    - Nicotiana L., genus widely distributed, with 52 American species, 23 Australian, and one African

Solanoideae

Capsicum frutescens cultivar "tabasco", a solanoidea

Flor de beleño (Hyoscyamus niger)

Nicandra physalodes flower

Solandra maxima flower

In the fruit of Physalis peruviana (tomatillo), the persistent calyx surrounds the fruit.

Iochroma australe flower

Jaltomata procumbens flower

Solanum bonariense flower

Flower of Solanum betaceum (=Cyphomandra betacea)

Acnistus arborescens flower

Scopolia carniolica flower

- Capsiceae Dumort
    - Capsicum L. includes 40 accepted neotropical species
    - Lycianthes (Dunal) Hassler, some 200 species distributed throughout America and Asia

- Datureae G. Don, two genera are perfectly differentiated at both the morphological and molecular levels, Brugmansia includes tree species, while Datura contains herbs or shrubs, the latter genus can be divided into three sections: Stramonium, Dutra and Ceratocaulis.
    - Brugmansia Persoon, six species from the Andes
    - Datura L., 12 neotropical species
- Hyoscyameae Endl.
    - Anisodus Link, four species from China, India and the Himalayas
    - Atropa L., four Euro-Asiatic species
    - Atropanthe Pascher, monotypic genus from China
    - Hyoscyamus L., 10 accepted species distributed from the Mediterranean to China
    - Physochlaina G. Don, 6 accepted Euro-Asiatic species
    - Przewalskia Maxim., one species from China
    - Scopolia Jacq., disjointed distribution with one European species and another from Japan
- Jaboroseae Miers
    - Jaborosa Juss., genus that includes 23 species from South America.
- Solandreae Miers
    - Subtribe Juanulloinae consists 10 genera of trees and epiphytic shrubs with a neotropical distribution . Some of these genera (Dyssochroma, Merinthopodium and Trianaea) show a clear dependency on various species of bats both for pollination and dispersion of seeds.
        - Dyssochroma Miers, two species from the south of Brazil
        - Ectozoma Miers
        - Hawkesiophyton Hunz.
        - Juanulloa Ruiz & Pav., 11 species from South and Central America
        - Markea Rich., 9 species from South and Central America
        - Merinthopodium J. Donn. Sm. three species originating from South America
        - Rahowardiana D' Arcy
        - Schultesianthus Hunz., eight neotropical species
        - Trianaea Planch. & Linden, six South American species
    - Subtribe Solandrinae, a monotypical subtribe, differs from Juanulloinae in that its embryos have incumbent cotyledons and semi-inferior ovaries.
    - Solandra Sw., 10 species from the neotropical regions of America

- Lycieae Hunz. has three genera of woody plants, which grow in arid or semiarid climates. The cosmopolitan genus Lycium is the oldest in the tribe and it has the greatest morphological variability.Molecular phylogenetic studies suggest both Grabowskia and Phrodus should be included in the Lycium, and this genus, along with Nolana and Sclerophylax, form a clade (Lyciina), which currently lacks a taxonomic category. The red fleshy berries dispersed by birds are the main type of fruit in Lycium. The different types of fruit in this genus have evolved from the type of berry just mentioned to a drupe with a reduced number of seeds.
    - Grabowskia Schltdl., three species from South America
    - Lycium L., 83 cosmopolitan species
    - Phrodus Miers, two species endemic to the north of Chile
- Mandragoreae (Wettst.) Hunz. & Barboza tribe does not have a defined systematic position according to molecular phylogenetic studies.
    - Mandragora L., two species from Eurasia
- Nicandreae Wettst. is a tribe with two South American genera. Molecular phylogenetic studies indicate the genera are not interrelated nor are they related with other genera of the family, so their taxonomic position is uncertain.
    - Exodeconus Raf., six species from western South America
    - Nicandra Adans, one species distributed throughout neotropical regions
- Nolaneae Rchb. are mostly herbs and small shrubs with succulent leaves, they have very beautiful flowers that range from white to various shades of blue, their fruit is schizocarpal, giving rise to various nuts.
    - Nolana L., 89 species distributed throughout western South America
- Physaleae Miers, is a large tribe that is the sister of Capsiceae.
    - Subtribe Iochrominae (Miers) Hunz., a clade within the Physaleae tribe. contains 37 species, mainly distributed in the Andes, assigned to six genera. The members of this subtribe are characterized by being woody shrubs or small trees with attractive tubular or rotated flowers. They also possess great floral diversity, containing every type is present in the family. Their flowers can be red, orange, yellow, green, blue, purple, or white. The corolla can be tubular to rotated, with a variation of up to eight times in the length of the tube between the various species.
        - Acnistus Schott, one species distributed throughout the neotropics
        - Dunalia Kunth., five species from the Andes
        - Iochroma Benth., 24 species from the Andes
        - Saracha Ruiz & Pav., two species from the Andes.
        - Vassobia Rusby, two South American species
        - Eriolarynx Hunz., three species from Argentina and Bolivia

- Physalinae (Miers) Hunz. , a monophyletic subtribe, contains 10 genera and includes herbs or woody shrubs with yellow, white, or purple solitary axillary flowers pollinated by bees. Once pollination occurs, the corolla falls and the calyx expands until it entirely covers the boll that is developing (the calyx is called accrescent). In many species, the calyx turns yellow or orange on maturity. The berries contain many greenish to yellow-orange seeds, often with red or purple highlights.
    - Brachistus Miers, three species from Mexico and Central America
    - Chamaesaracha (A.Gray) Benth. & Hook., has 10 species from Mexico and Central America.
    - Leucophysalis Rydberg, includes 3 species from the south west of the United States and Mexico.
    - Margaranthus Schlecht., with 1 species from Mexico.
    - Oryctes S. Watson, monotypic genus from the south west of the United States.
    - Quincula Raf. with just 1 species from the south west of the United States and from Mexico.
    - Physalis L., the largest genus of the subtribe, with 85 species distributed through the tropical regions of the Americas and with 1 species in China.
    - Witheringia L' Heritier, genus with 15 species from neotropical regions.
    - Tzeltalia, genus segregated from Physalis, with 2 species distributed throughout Mexico and Guatemala.
    - Darcyanthus, genus with just 1 species originating in Bolivia and Peru.
- Subtribe Salpichroinae, this is a subtribe of Physaleae that includes 16 American species distributed in 1 genera:
    - Nectouxia Kunth., monotypic genus that is endemic to Mexico.
    - Salpichroa Miers, genus with 15 species from the Andes and other regions of South America.
- Subtribe Withaninae, is a subtribe of Physaleae with a broad distribution, including 9 genera:
    - Archiphysalis Kuang, with 3 species from China and Japan.
    - Athenaea Sendtn., which includes 7 species from Brazil.
    - Aureliana Sendtn., with 5 species from South America.

- Melissia Hook. f., monotypic genus from Santa Elena with the common name St. Elena boxwood.
- Physalisastrum Makino, with 9 Asiatic species.
- Tubocapsicum (Wettst.) Makino, with just one species endemic to China.
- Withania Pauq., with 10 species native to the Canary Islands, Africa and Nepal.
- Cuatresia Hunz., with 11 neotropical species. Molecular studies indicate that this genus, along with Deprea and Larnax has an uncertain taxonomic position.
- Deprea Raf., with 6 neotropical species.
- Larnax Miers, many taxonomists consider it to be a synonym for Deprea, contains 22 species native to the Andes.

- Tribe Solaneae. The genera Cyphomandra Sendtn., Discopodium Hochst. , Normania Lowe, Triguera Cav.and Lycopersicum Mill have been transferred to Solanum. The subtribe is therefore composed of two genera:
  - Jaltomata Schltdl., which contains 50 neotropical species.
  - Solanum L., the largest genus in the family and one of the broadest of the angiosperms, with 1,328 species distributed across the whole world.
- Genera with doubtful taxonomic positions (Incertae sedis)

The following genera have still not been placed in any of the recognized subfamilies within the solanaceas.

- Duckeodendron Kuhlmannb, monotypic genus from the Amazon rainforest.
- Parabouchetia Baillon, poorly-known,monotypic genus from Brazil.
- Pauia Deb. & Dutta, monotypic genus from Assam and Arunachal Pradesh in N.E.India

Genera and Distribution of Species

The Solanaceae contain 98 genera and some 2,700 species. Despite this immense richness of species, they are not uniformly distributed between the genera. The eight most important genera contain more than 60% of the species, as shown in the table below. Solanum – the genus that typifies the family - includes nearly 50% of the total species of the solanaceas.

Flowers and foliage of Cestrum parqui.

Genera	Approximate number of species
Solanum	1,330
Lycianthes	200
Cestrum	150
Nolana	89
Physalis	85
Lycium	85
Nicotiana	76
Brunfelsia	45
Estimated number of species in the family	2,700

Economic Importance

The solanaceas include such important food species as the potato (Solanum tuberosum), the tomato (Solanum lycopersicum), the pepper (Capsicum annuum) and the aubergine or egg plant (Solanum melongena). Nicotiana tabacum, originally from South America, is now cultivated throughout the world to produce tobacco. Many solanaceas are important weeds in various parts of the world. Their importance lies in the fact that they can host pathogens or diseases of the cultivated plants, therefore their presence increases the loss of yield or the quality of the harvested product. An example of this can be seen with Acnistus arborescens and Browalia americana that host thrips, which cause damage to associated cultivated plants, and certain species of Datura that play host to various types of virus that are later transmitted to cultivated solanaceas. Some species of weeds such as, for example Solanum mauritianum in South Africa represent such serious ecological and economic problems that studies are being carried out with the objective of developing a biological control through the use of insects.

Pink flower of the Brugmansia

Petunia hybrida, a herbaceous annual that is commonly used in gardens

Various solanaceas species are grown as ornamental trees or shrubs. Examples include Brugmansia x candida ("Angel's Trumpet") grown for its large pendulous trumpet-shaped flowers, or Brunfelsia latifolia, whose flowers are very fragrant and change colour from violet to white over a period of 3 days. Other shrub species that are grown for their attractive flowers are Lycianthes rantonnetii (Blue Potato Bush or Paraguay Nightshade) with violet-blue flowers and Nicotiana glauca ("Tree Tobacco") Other solanacea species and genera that are grown as ornamentals are the petunia (Petunia × hybrida), Lycium, Solanum, Cestrum,Calibrachoa × hybrida and Solandra. There is even a hybrid between Petunia and Calibrachoa (which constitutes a new nothogenus called × Petchoa G. Boker & J. Shaw) that is being sold as an ornamental. Many other species, in particular those that produce alkaloids, are used in pharmacology and medicine (Nicotiana, Hyoscyamus, and Datura).

Solanaceas and the Genome

Many of the species belonging to this family, among them tobacco and the tomato, are model organisms that are used for research into fundamental biological questions. One of the aspects of the solanaceas' genomics is an international project that is trying to understand how the same collection of genes and proteins can give rise to a group of organisms that are so morphologically and ecologically different. The first objective of this project was to sequence the genome of the tomato. In order to achieve this each of the 12 chromosomes of the tomato's haploid genome was assigned to different sequencing centres in different countries. So chromosomes 1 and 10 were sequenced in the United States, 3 and 11 in China, 2 in Korea, 4 in Britain, 5 in India, 7 in France, 8 in Japan, 9 in Spain and 12 in Italy. The sequencing of the mitochondrial genome was carried out in Argentina and the chloroplast genome was sequenced in the European Union.

TOMATO

The tomato (see pronunciation) is the edible, red fruit of Solanum lycopersicum, commonly known as a **tomato plant**, which belongs to the nightshade family, Solanaceae.

The species originated in Central and South America. The Nahuatl (Aztec language) word tomatl gave rise to the Spanish word "tomate", from which the English word tomato originates.

Cross-section and full view of a hothouse (greenhouse-grown) tomato

Numerous varieties of tomato are widely grown in temperate climates across the world, with greenhouses allowing its production throughout the year and in cooler areas. The plants typically grow to 1–3 meters (3–10 ft) in height and have a weak stem that often sprawls over the ground and vines over other plants. It is a perennial in its native habitat, and grown as an annual in temperate climates. An average common tomato weighs approximately 100 grams (4 oz).

Its use as a food originated in Mexico, and spread throughout the world following the Spanish colonization of the Americas. Tomato is consumed in diverse ways, including raw, as an ingredient in many dishes, sauces, salads, and drinks. While tomatoes are botanically berry-type fruits, they are considered culinary vegetables, being ingredients of savory meals.

Names

Etymology

The word "tomato" comes from the Spanish tomate, which in turn comes from the Nahuatl word tomatl. It first appeared in print in 1595.

The native Mexican tomatillo is tomate (in Nahuatl: tomātl, meaning "fat water" or "fat thing"). When Aztecs started to cultivate the Andean fruit, bigger and red, they called the new species xitomatl (or jitomates) (pronounced), ("plump thing with navel" or "fat water with navel"). After their conquest of Tenochtitlan, Spaniards exported tomatoes (jitomates) to the rest of the world with the name tomate, so numerous languages use forms of the word "tomato" (tomate) to refer to the red tomato instead of the green tomatillo. It first appeared in print in 1595. The scientific species epithet lycopersicum means "wolf peach", and comes from German werewolf myths. These legends said that deadly nightshade was used by witches and sorcerers in potions to transform themselves into werewolves, so the tomato's similar, but much larger, fruit was called the "wolf peach" when it arrived in Europe.

Derivations

The Italian word, pomodoro (from pomo d'oro "apple of gold") was borrowed into Polish, and via Russian, into several other languages. Similarly, the now rare German term Paradeisapfel, which means "apple of paradise", is still heard in the form paradeiser in the Bavarian and Austrian dialects, and was borrowed into modern Hungarian, Slovenian and Serbian.

Pronunciation

The usual pronunciations of "tomato" are (usual in American English) and (usual in British English). The word's dual pronunciations were immortalized in Ira and George Gershwin's 1937 song "Let's Call the Whole Thing Off" ("You like and I like / You like and I like and have become a symbol for nitpicking pronunciation disputes. In this capacity, it has even become an American and British slang term: saying when presented with two choices can mean “What’s the difference?” or “It’s all the same to me”.

Description

Tomato plants are vines, initially decumbent, typically growing 180 cm (6 ft) or more above the ground if supported, although erect bush varieties have been bred, generally 100 cm (3 ft) tall or shorter. Indeterminate types are “tender” perennials, dying annually in temperate climates (they are originally native to tropical highlands), although they can live up to three years in a greenhouse in some cases. Determinate types are annual in all climates.

Tomato plants are dicots, and grow as a series of branching stems, with a terminal bud at the tip that does the actual growing. When that tip eventually stops growing, whether because of pruning or flowering, lateral buds take over and grow into other, fully functional, vines.

Tomato flower

Tomato vines are typically pubescent, meaning covered with fine short hairs. These hairs facilitate the vining process, turning into roots wherever the plant is in contact with the ground and moisture, especially if the vine’s connection to its original root has been damaged or severed.

Most tomato plants have compound leaves, and are called regular leaf (RL) plants, but some cultivars have simple leaves known as potato leaf (PL) style because of their resemblance to that particular relative. Of RL plants, there are variations, such as rugose leaves, which are deeply grooved, and variegated, angora leaves, which have additional colors where a genetic mutation causes chlorophyll to be excluded from some portions of the leaves.

The leaves are 10–25 cm (4–10 in) long, odd pinnate, with five to 9 leaflets on petioles, each leaflet up to 8 cm (3 in) long, with a serrated margin; both the stem and leaves are densely glandular-hairy.

Their flowers, appearing on the apical meristem, have the anthers fused along the edges, forming a column surrounding the pistil's style. Flowers in domestic cultivars tend to be self-fertilizing. The flowers are 1–2 cm (0.4–0.8 in) across, yellow, with five pointed lobes on the corolla; they are borne in a cyme of three to 12 together.

Tomato fruit is classified as a berry. As a true fruit, it develops from the ovary of the plant after fertilization, its flesh comprising the pericarp walls. The fruit contains hollow spaces full of seeds and moisture, called locular cavities. These vary, among cultivated species, according to type. Some smaller varieties have two cavities, globe-shaped varieties typically have three to five, beefsteak tomatoes have a great number of smaller cavities, while paste tomatoes have very few, very small cavities.

For propagation, the seeds need to come from a mature fruit, and be dried or fermented before germination.

Classification

In 1753, Linnaeus placed the tomato in the genus Solanum (alongside the potato) as Solanum lycopersicum. In 1768, Philip Miller moved it to its own genus, naming it Lycopersicon esculentum. This name came into wide use, but was in breach of the plant naming rules. Although the name Lycopersicon lycopersicum may be found, it is not used because it violates the International Code of Nomenclature barring the use of tautonyms in botanical nomenclature.

Genetic evidence has now shown that Linnaeus was correct to put the tomato in the genus Solanum, making Solanum lycopersicum the correct name. Both names, however, will probably be found in the literature for some time. Two of the major reasons some still consider the genera separate are the leaf structure (tomato leaves are markedly different from any other Solanum), and the biochemistry (many of the alkaloids common to other Solanum species are conspicuously absent in the tomato). Hybrids of tomato and diploid potato can be created in the lab by somatic fusion, and are partially fertile, providing evidence of the close relationship between these species.

Wild Species

Including Solanum lycopersicum, there are currently 13 species recognized in Solanum section Lycopersicon. Three of these species—S. cheesmaniae, S. galapagense, and S. pimpinellifolium—are fully cross compatible with domestic tomato. Four more species—S. chmielewskii, S. habrochaites, S. neorickii, and S. pennelli—can be readily crossed with domestic tomato, with some limitations. Five species—S. arcanum, S. chilense, S. corneliomulleri, S. huaylasense, and S. peruvianum—can be crossed with domestic tomato with difficulty and usually require embryo rescue to produce viable plants. The Lycopersicon section has not been fully sampled within wild species in the South American range, so new species may be added in the future.

Solanum section Lycopersicoides and section Juglandifolium are represented by two species each that are considered bridge species genetically intermediate between tomato and non-tuber bearing potato species. S. lycopersicoides can be crossed with domestic tomato and introgression lines have been developed. This species was significant in moving the domestic tomato from separate genus status into the Solanum group because it directly links the tomato into the potato family.

Genetic Modification

Tomatoes that have been modified using genetic engineering have been developed, and although none are commercially available now, they have been in the past. The first commercially available genetically modified food was a variety of tomato named the Flavr Savr, which was engineered to have a longer shelf life. Scientists are continuing to develop tomatoes with new traits not found in natural crops, such as increased resistance to pests or environmental stresses. Other projects aim to enrich tomatoes with substances that may offer health benefits or provide better nutrition.

Research on tomatoes

An international consortium of researchers from 10 countries, among them researchers from the Boyce Thompson Institute for Plant Research, began sequencing the tomato genome in 2004, and is creating a database of genomic sequences and information on the tomato and related plants. A prerelease version of the genome was made available in December 2009. The genomes of its mitochondria and chloroplasts are also being sequenced as part of the project. The complete genome for the cultivar Heinz 1706 was published on 31 May 2012 in Nature.Since many other fruits, like strawberries, apples, melons, and bananas share the same characteristics and genes, researchers stated the published genome could help to improve food quality, food security and reduce costs of all of these fruits.

Breeding

Breeding programs are ongoing by individuals, universities, corporations, and organizations. The Tomato Genetic Resource Center, Germplasm Resources Information Network, AVRDC, and numerous seed banks around the world store seed representing genetic variations of value to modern agriculture. These seed stocks are available for legitimate breeding and research efforts. While individual breeding efforts can produce useful results, the bulk of tomato breeding work is at universities and major agriculture-related corporations. These efforts have resulted in significant regionally adapted breeding lines and hybrids, such as the Mountain series from North Carolina. Corporations including Heinz, Monsanto, BHNSeed, Bejoseed, etc., have breeding programs that attempt to improve production, size, shape, color, flavor, disease tolerance, pest tolerance, nutritional value, and numerous other traits.

Fruit versus Vegetable

Botanically, a tomato is a fruit, a berry, consisting of the ovary, together with its seeds, of a flowering plant. However, the tomato has a much lower sugar content than other edible fruits, and is therefore not as sweet. Typically served as part of a salad or main course of a meal, rather than at dessert, it is considered a "culinary vegetable". One exception is that tomatoes are treated as a fruit in home canning practices: they are acidic enough to process in a water bath rather than a pressure cooker as vegetables require. Tomatoes are not the only food source with this ambiguity: bell peppers, cucumbers, green beans, eggplants, avocados, and squashes of all kinds (such as zucchini and pumpkins) are all botanically fruits, yet cooked as vegetables. This has led to legal dispute in the United States. In 1887, U.S. tariff laws that imposed a duty on vegetables, but not on fruits, caused the tomato's status to become a matter of legal importance. The U.S. Supreme Court settled this controversy on 10 May 1893, by declaring that the tomato is a vegetable, based on the popular definition that classifies vegetables by use, that they are generally served with dinner and not dessert (Nix v. Hedden (149 U.S. 304)). The holding of this case applies only to the interpretation of the Tariff Act of 3 March 1883, and the court did not purport to reclassify the tomato for botanical or other purposes.

Cultivation

The tomato is grown worldwide for its edible fruits, with thousands of cultivars. A fertilizer with an NPK ratio of 5-10-10 is often sold as tomato fertilizer or vegetable fertilizer, although manure and compost are also used.

Tomato plants 7 days after planting

27 days after planting

52-day-old plant, first fruits

Production

In 2013, world production of tomatoes was 163.4 million tonnes, with China accounting for 31% of the total, followed by India, the United States and Turkey (table).

In 2012, tomato production was valued at 59 billion dollars, the eighth most valuable agricultural product worldwide.

Rank	**Country**	**Production** **millions of tonnes**
1	China	50.6
2	India	18.3
3	United States	12.6
4	Turkey	11.8
5	Egypt	8.5
6	Iran	6.2
7	Italy	5.6
8	Spain	4.9
9	Portugal	1.4
10	Greece	1.0
	World	**163.4**

Tomato seedlings growing indoors

Green tomatoes nestled on the vine

Varieties

There are around 7,500 tomato varieties grown for various purposes having been selected with varying fruit types, and for optimum growth in differing growing conditions.

Tomato varieties can be divided into categories based on shape and size.

- Beefsteak tomatoes are 10 cm (4 in) or more in diameter, often used for sandwiches and similar applications. Their kidney-bean shape, thinner skin, and shorter shelf life makes commercial use impractical.

- Plum tomatoes, or paste tomatoes (including pear tomatoes), are bred with a lower water /higher solids content for use in tomato sauce and paste, for canning and sauces and are usually oblong 7–9 cm (3–4 in) long and 4–5 cm (1.6–2.0 in) diameter; like the Roma-type tomatoes, important cultivars in the Sacramento Valley.
- Cherry tomatoes are small and round, often sweet tomatoes, about the same 1–2 cm (0.4–0.8 in) size as the wild tomato.
- Grape tomatoes are smaller and oblong, a variation on plum tomatoes.
- Campari tomatoes are sweet and noted for their juiciness, low acidity, and lack of mealiness, bigger than cherry tomatoes, and smaller than plum tomatoes.
- Tomberries, tiny tomatoes, about 5 mm in diameter
- Oxheart tomatoes can range in size up to beefsteaks, and are shaped like large strawberries.
- Pear tomatoes are pear-shaped and can be based upon the San Marzano types for a richer gourmet paste.
- "Slicing" or "globe" tomatoes are the usual tomatoes of commerce, used for a wide variety of processing and fresh eating. The most widely grown commercial tomatoes tend to be in the 5–6 cm (2.0–2.4 in) diameter range.

Tomatoes are also classified as determinate or indeterminate. Determinate, or bush, types bear a full crop all at once and top off at a specific height; they are often good choices for container growing. Determinate types are preferred by commercial growers who wish to harvest a whole field at one time, or home growers interested in canning. Indeterminate varieties develop into vines that never top off and continue producing until killed by frost. They are preferred by home growers and local-market farmers who want ripe fruit throughout the season. As an intermediate form, there are plants sometimes known as vigorous determinate or semi-determinate; these top off like determinates, but produce a second crop after the initial crop. The majority of heirloom tomatoes are indeterminate, although some determinate heirlooms exist.

Early tomatoes and cool-summer tomatoes bear fruit even where nights are cool, which usually discourages fruit set. There are varieties high in beta carotenes and vitamin A, hollow tomatoes and tomatoes that keep for months in storage. In 1973, Israeli scientists developed the world's first long shelf-life commercial tomato varieties.

Heirloom tomatoes are becoming increasingly popular, particularly among home gardeners and organic producers, since they tend to produce more interesting and flavorful crops at the cost of disease resistance and productivity. The definition of an heirloom tomato is vague, but unlike commercial hybrids, all are self-pollinators that have bred true for 40 years or more. Quite a few seed merchants and banks provide a large selection of heirloom seeds. Home cultivars are often bred for flavor to the exclusion of all other qualities, while commercial cultivars are bred for factors like consistent size and shape, disease and pest resistance, suitability for mechanized picking and shipping, and ability to ripen after picking. Hybrid plants remain common, since they tend to be heavier producers, and sometimes combine unusual characteristics of heirloom tomatoes with the ruggedness of conventional commercial tomatoes.

Most modern tomato cultivars are smooth surfaced, but some older tomato cultivars and most modern beefsteaks show pronounced ribbing, a feature that may have been common to virtually all pre-Columbian cultivars. While virtually all commercial tomato varieties are red, some cultivars – especially heirlooms – produce fruit in green, yellow, orange, pink, black, brown, ivory, white, and purple. Such fruits are not widely available in grocery stores, nor are their seedlings available in typical nurseries, but they can be bought as seed. Variations include multicolored fruit with stripes (Green Zebra), fuzzy skin on the fruit (Fuzzy Peach, Red Boar), multiple colors (Hillbilly, Burracker's Favorite, Lucky Cross), etc.

Various heirloom tomato cultivars

A variety of specific cultivars, including Brandywine (biggest red), Black Krim (lower left) and Green Zebra (top left)

Diseases and Pests

Tomato cultivars vary widely in their resistance to disease. Modern hybrids focus on improving disease resistance over the heirloom plants.

Various forms of mildew and blight are common tomato afflictions, which is why tomato cultivars are often marked with a combination of letters that refer to specific disease resistance. The most common letters are: **LB** - late blight, **V** – verticillium wilt, **F** – fusarium wilt strain I, **FF** – fusarium wilt strain I and II, **N** – nematodes, **T** – tobacco mosaic virus, and **A** – alternaria.

Some common tomato pests are stink bugs, cutworms, tomato hornworms and tobacco hornworms, aphids, cabbage loopers, whiteflies, tomato fruitworms, flea beetles, red spider mite, slugs, and Colorado potato beetles. The tomato russet mite, Aculops lycopersici, feeds on foliage and young fruit of tomato plants, causing shrivelling and necrosis of leaves, flowers, and fruit, possibly killing the plant.

Tomato fruitworm feeding on unripe tomato

A common tomato disease is tobacco mosaic virus. Handling cigarettes and other infected tobacco products can transmit the virus to tomato plants.

Another particularly dreaded disease is curly top, carried by the beet leafhopper, which interrupts the lifecycle. As the name implies, it has the symptom of making the top leaves of the plant wrinkle up and grow abnormally.

A split heirloom tomato, caused fluctuation in water availability

After an insect attack tomato plants produce systemin, a plant peptide hormone . Systemin activates defensive mechanisms, such as the production of protease inhibitors to slow the growth of insects. The hormone was first identified in tomatoes, but similar proteins have been identified in other species since.

Companion Plants

Tomatoes serve, or are served by, a large variety of companion plants.

Among the most famous pairings is the tomato plant and carrots; studies supporting this relationship have produced a popular book about companion planting, Carrots Love Tomatoes.

The devastating tomato hornworm has a major predator in various parasitic wasps, whose larvae devour the hornworm, but whose adult form drinks nectar from tiny-flowered plants like umbellifers. Several species of umbellifer are therefore often grown with tomato plants, including parsley, queen anne's lace, and occasionally dill. These also attract predatory flies that attack various tomato pests.

Borage is thought to repel the tomato hornworm moth.

Plants with strong scents, like alliums (onions, chives, garlic) and mints (basil, oregano, spearmint) are thought to mask the scent of the tomato plant, making it harder for pests to locate it, or to provide an alternative landing point, reducing the odds of the pests from attacking the correct plant. These plants may also subtly impact the flavor of tomato fruit.

Ground cover plants, including mints, stabilize moisture loss around tomato plants and other solaneae, which come from very humid climates, and therefore may prevent moisture-related problems like blossom end rot.

Tap-root plants like dandelions break up dense soil and bring nutrients from down below a tomato plant's reach, possibly benefiting their companion.

Tomato plants can protect asparagus from asparagus beetles, because they contain solanum that kills this pest, while asparagus plants (as well as marigolds) contain a chemical that repels root nematodes known to attack tomato plants.

Pollination

In the wild, original state, tomatoes required cross-pollination; they were much more self-incompatible than domestic cultivars. As a floral device to reduce selfing, the pistil of wild tomatoes extends farther out of the flower than today's cultivars. The stamens were, and remain, entirely within the closed corolla.

Tomato flower in full bloom, associated with a young, developing fruit.

The flower and leaves are visible in this photo of a tomato plant.

As tomatoes were moved from their native areas, their traditional pollinators, (probably a species of halictid bee) did not move with them. The trait of self-fertility became an advantage, and domestic cultivars of tomato have been selected to maximize this trait.

This is not the same as self-pollination, despite the common claim that tomatoes do so. That tomatoes pollinate themselves poorly without outside aid is clearly shown in greenhouse situations, where pollination must be aided by artificial wind, vibration of the plants (one brand of vibrator is a wand called an "electric bee" that is used manually), or more often today, by cultured bumblebees. The anther of a tomato flower is shaped like a hollow tube, with the pollen produced within the structure, rather than on the surface, as in most species. The pollen moves through pores in the anther, but very little pollen is shed without some kind of outside motion. The best source of outside motion is a sonicating bee, such as a bumblebee, or the original wild halictid pollinator. In an outside setting, wind or animals provide sufficient motion to produce commercially viable crops.

Fruit Formation

Pollination and fruit formation depend on meiosis. Meiosis is central to the processes by which diploid microspore mother cells within the anther give rise to haploid pollen grains, and megaspore mother cells in ovules that are contained within the ovary give rise to haploid nuclei. Union of haploid nuclei from pollen and ovule (fertilization) can occur either by self- or cross-pollination. Fertilization leads to the formation of a diploid zygote that can then develop into an embryo within the emerging seed. Repeated fertilizations within the ovary are accompanied by maturation of the ovary to form the tomato fruit.

Homologs of the recA gene, including rad51, play a key role in homologous recombinational repair of DNA during meiosis. A rad51 homolog is present in the anther of tomato (Lycopersicon esculentum), suggesting that recombinational repair occurs during meiosis in tomato.

Hydroponic and Greenhouse Cultivation

Tomatoes are often grown in greenhouses in cooler climates, and there are cultivars such as the British 'Moneymaker' and a number of cultivars grown in Siberia that are specifically bred for indoor growing. In more temperate climates, it is not uncommon to start seeds in greenhouses during the late winter for future transplant.

Greenhouse tomato production in large-acreage commercial greenhouses and owner-operator stand-alone or multiple-bay greenhouses is on the increase, providing fruit during those times of the year when field-grown fruit is not readily available. Smaller sized fruit (cherry and grape), or cluster tomatoes (fruit-on-the-vine) are the fruit of choice for the large commercial greenhouse operators while the beefsteak varieties are the choice of owner-operator growers.

Hydroponic technique is often used in hostile growing environments, as well as high-density plantings.

Picking and Ripening

To facilitate transportation and storage, tomatoes are often picked unripe (green) and ripened in storage with ethylene. Unripe tomatoes are firm. As they ripen they soften until reaching the ripe state where they are red or orange in color and slightly soft to the touch. Ethylene is a hydrocarbon gas produced by many fruits that acts as the molecular cue to begin the ripening process. Tomatoes ripened in this way tend to keep longer, but have poorer flavor and a mealier, starchier texture than tomatoes ripened on the plant. They may be recognized by their color, which is more pink or orange than the other ripe tomatoes' deep red, depending on variety.

A machine-harvestable variety of tomato (the "square tomato") was developed in the 1950s by University of California, Davis's Gordie C. Hanna, which, in combination with the development of a suitable harvester, revolutionized the tomato-growing industry. This type of tomato is grown commercially near plants that process and can tomatoes, tomato sauce, and tomato paste. They are harvested when ripe and are flavorful when picked. They are harvested 24 hours a day, seven days a week during a 12 to 14 week season, and immediately transported to packing plants, which operate on the same schedule. California is a center of this sort of commercial tomato production and produces about a third of the processed tomatoes produced in the world.

In 1994, Calgene introduced a genetically modified tomato called the FlavrSavr, which could be vine ripened without compromising shelf life. However, the product was not commercially successful, and was sold only until 1997. Slow-ripening cultivars of tomato have been developed by crossing a non-ripening cultivar with ordinary cultivars.

Yield

The world dedicated 4.8 million hectares in 2012 for tomato cultivation and the total production was about 161.8 million tonnes. The average world farm yield for tomato was 33.6 tonnes per hectare, in 2012.

Tomato farms in the Netherlands were the most productive in 2012, with a nationwide average of 476 tonnes per hectare, followed by Belgium (463 tonnes per hectare) and Iceland (429 tonnes per hectare).

Records

The heaviest tomato ever, weighing 3.51 kg (7 lb 12 oz), was of the cultivar "Delicious", grown by Gordon Graham of Edmond, Oklahoma in 1986. The largest tomato plant grown was of the cultivar "Sungold" and reached 19.8 m (65 ft) in length, grown by Nutriculture Ltd (UK) of Mawdesley, Lancashire, UK, in 2000.

The "tomato tree" as seen by guests on the Living with the Land boat ride at Epcot, Lake Buena Vista, Florida

A massive "tomato tree" growing inside the Walt Disney World Resort's experimental greenhouses in Lake Buena Vista, Florida may have been the largest single tomato plant in the world. The plant has been recognized as a Guinness World Record Holder, with a harvest of more than 32,000 tomatoes and a total weight of 522 kg (1,151 lb). It yielded thousands of tomatoes at one time from a single vine. Yong Huang, Epcot's manager of agricultural science, discovered the unique plant in Beijing, China. Huang brought its seeds to Epcot and created the specialized greenhouse for the fruit to grow. The vine grew golf ball-sized tomatoes, which were served at Walt Disney World restaurants. The tree developed a disease and was removed in April 2010 after about 13 months of life.

Consumption

The tomato is now grown and eaten around the world. It is used in diverse ways, including raw in salads, and processed into ketchup or tomato soup. Unripe green tomatoes can also be breaded and fried, used to make salsa, or pickled. Tomato juice is sold as a drink, and is used in cocktails such as the Bloody Mary.

Stuffed tomatoes (stuffed with hard-boiled egg and Parmesan)

Bruschetta, an antipasto dish

Tomatoes are acidic, making them especially easy to preserve in home canning whole, in pieces, as tomato sauce or paste. The fruit is also preserved by drying, often in the sun, and sold either in bags or in jars with oil.

Although tomatoes originated in the Americas, they have become extensively used in Mediterranean cuisine. They are a key ingredient in pizza, and are commonly used in pasta sauces. They are also used in gazpacho (Spanish cuisine) and pa amb tomàquet (Catalan cuisine).

Though it is botanically a berry, a subset of fruit, the tomato is a vegetable for culinary purposes, because of its savory flavor (see below).

Storage

Tomatoes keep best unwashed at room temperature and out of direct sunlight. It is not recommended to refrigerate them as this can harm the flavor. Tomatoes stored cold tend to lose their flavor permanently.

Storing stem down can prolong shelf life, as it may keep from rotting too quickly.

Tomatoes that are not yet ripe can be kept in a paper bag till ripening.

Safety

Plant Toxicity

Leaves, stems, and green unripe fruit of the tomato plant contain small amounts of the toxic alkaloid tomatine. They also contain solanine, a toxic alkaloid found in potato leaves and other plants in the nightshade family. Use of tomato leaves in herbal tea has been responsible for at least one death. However, levels of tomatine in foliage and green fruit are generally too small to be dangerous unless large amounts are consumed, for example, as greens. Small amounts of tomato foliage are sometimes used for flavoring without ill effect, and the green fruit is sometimes used for cooking, particularly as fried green tomatoes. Compared to potatoes the amount of solanine in green or ripe tomatoes is low; however, even in the case of potatoes while solanine poisoning resulting from dosages several times normal human consumption has been demonstrated, actual cases of poisoning resulting from excessive consumption of potatoes that have high concentration of solanine are rare.

Tomato plants can be toxic to dogs if they eat large amounts of the fruit, or chew plant material.

Salmonella

Tomatoes were linked to seven salmonella outbreaks between 1990 and 2005, and may have been the cause of a salmonellosis outbreak causing 172 illnesses in 18 U.S. states in 2006, The 2008 United States salmonellosis outbreak also caused the removal of tomatoes from stores and restaurants across the United States and parts of Canada, although other foods, including jalapeño and serrano peppers, may have been involved.

A sign posted at a Havelock, North Carolina Burger King tells customers that no tomatoes are available due to the 2008 United States salmonellosis outbreak.

Nutrition

Red tomatoes, raw	
Nutritional value per 100 g (3.5 oz)	
Energy	74 kJ (18 kcal)
Carbohydrates	3.9 g
Sugars	2.6 g
Dietary fiber	1.2 g
Fat	0.2 g
Protein	0.9 g
Vitamins	
Vitamin A equiv. beta-carotene lutein zeaxanthin	(5%) 42 μg (4%) 449 μg 123 μg
Thiamine (B1)	(3%) 0.037 mg
Niacin (B3)	(4%) 0.594 mg
Vitamin B6	(6%) 0.08 mg
Vitamin C	(17%) 14 mg
Vitamin E	(4%) 0.54 mg
Vitamin K	(8%) 7.9 μg
Minerals	

Magnesium	(3%) 11 mg
Manganese	(5%) 0.114 mg
Phosphorus	(3%) 24 mg
Potassium	(5%) 237 mg
Other constituents	
Water	94.5 g
Lycopene	2573 µg
Link to USDA Database entry	
• Units • µg = micrograms • mg = milligrams • IU = International units	
Percentages are roughly approximated using US recommendations for adults. Source: USDA Nutrient Database	

A tomato is 95% water, contains 4% carbohydrates and less than 1% each of fat and protein (table). In a 100 gram amount, raw tomatoes supply 18 calories and are a moderate source of vitamin C (17% of the Daily Value), but otherwise are absent of significant nutrient content (table).

Potential Health Effects

There is no conclusive evidence that the lycopene in tomatoes or in supplements affects the onset of cardiovascular diseases or cancer.

In the United States, supposed health benefits of consuming tomatoes, tomato products or lycopene to affect cancer cannot be mentioned on packaged food products without a qualified health claim statement. In a scientific review of potential claims for lycopene favorably affecting DNA, skin exposed to ultraviolet radiation, heart function and vision, the European Food Safety Authority concluded there was insufficient evidence for lycopene having any of these effects.

In Popular Culture

On 30 August 2007, 40,000 Spaniards gathered in Buñol to throw 115,000 kg (254,000 lb) of tomatoes at each other in the yearly Tomatina festival.

In Ontario, Canada, member of provincial parliament Mike Colle named the tomato the official vegetable of the province and designated July 15 as Tomato Day under the Tomato Act in 2016.

Tomatoes have been designated the state vegetable of New Jersey. Arkansas took both sides by declaring the South Arkansas Vine Ripe Pink Tomato both the state fruit and the state vegetable in the same law, citing both its culinary and botanical classifications. In 2009, the state of Ohio passed a law making the tomato the state's official fruit. Tomato juice has been the official beverage of Ohio since 1965. A.W. Livingston, of Reynoldsburg, Ohio, played a large part in popularizing the tomato in the late 19th century; his efforts are commemorated in Reynoldsburg with an annual Tomato Festival.

Flavr Savr was the first commercially grown genetically engineered food licensed for human consumption.

The town of Buñol, Spain, annually celebrates La Tomatina, a festival centered on an enormous tomato fight. Tomatoes are a popular "nonlethal" throwing weapon in mass protests, and there was a common tradition of throwing rotten tomatoes at bad performers on a stage during the 19th century; today this is usually referenced as a metaphor. Embracing it for this protest connotation, the Dutch Socialist party adopted the tomato as their logo.

The US city of Reynoldsburg, Ohio calls itself "The Birthplace of the Tomato", claiming the first commercial variety of tomato was bred there in the 19th century.

Several US states have adopted the tomato as a state fruit or vegetable (see above).

EGGPLANT

Eggplant (Solanum melongena), or aubergine, is a species of nightshade grown for its edible fruit. Eggplant is the common name in North America and Australia, but British English uses aubergine. It is known in South Asia, Southeast Asia and South Africa as brinjal. Other common names are melongene, garden egg or guinea squash.

The fruit is widely used in cooking. As a member of the genus Solanum, it is related to the tomato and the potato. It was originally domesticated from the wild nightshade species, the thorn or bitter apple, S. incanum, probably with two independent domestications, one in South Asia and one in East Asia.

Description

The eggplant is a delicate, tropical perennial often cultivated as a tender or half-hardy annual in temperate climates. The stem is often spiny. The flower is white to purple, with a five-lobed corolla and yellow stamens. The egg-shaped glossy purple fruit has white flesh with a meaty texture. The cut surface of the flesh rapidly turns brown when the fruit is cut open.

It grows 40 to 150 cm (1.3 to 4.9 ft) tall, with large, coarsely lobed leaves that are 10 to 20 cm (3.9 to 7.9 in) long and 5 to 10 cm (2.0 to 3.9 in) broad. Semi-wild types can grow much larger, to 225 cm (7.38 ft) with large leaves over 30 cm (12 in) long and 15 cm (5.9 in) broad. On wild plants, the fruit is less than 3 cm (1.2 in) in diameter, but much larger in cultivated forms: 30 cm (12 in) or more in length.

Botanically classified as a berry, the fruit contains numerous small, soft seeds that, though edible, taste bitter because they contain nicotinoidalkaloids like the related tobacco.

Etymology and Regional Names

The plant and fruit have a profusion of English names: eggplant (North America, Australia), aubergine (Britain), brinjal (South Asia, South Africa, Malaysia, Singapore, terong (Indonesia), and West Indies, Trinidad), melongene (Caribbean), and formerly melongena and mad-apple.

Closeup of an eggplant flower of a long-fruited Chinese variety in Hong Kong

The word “eggplant” was first recorded in 1767, and was originally applied to white varieties; some 18th-century European cultivars were small, round, yellow or white, resembling goose or hen’s eggs. The other names, even mad-apple, all ultimately derive from a Dravidian word with reflexes in modern Malayalamvarutina and Tamil varutunai, transmitted through Sanskrit vātigama, Prakrit vāimana, Persian bādingān, and Arabic bādinjān .

The Anglo-Indian name “brinjal” or brinjaul comes from the Portuguese bringella, bringiela, or beringela,whereas the name baingan or baigan, also sometimes used in English in South Asia as well as in Trinidad, appears to be re-borrowed from the Sanskrit or Persian name.

Eggplant with chicken eggs

The Arabic name is the common source of all the European names for this plant, but through two distinct paths of transmission, with the melongene family coming through the eastern Mediterranean, and the aubergine family through the western Mediterranean. In the western Mediterranean, the Arabic (al)-bādinjān was borrowed as Spanish alberengena and berenjena, Catalan albergínia, and Portuguese beringela, whence the modern French aubergine (and the earlier albergine, albergaine, albergame, belingèle), the source of the British English aubergine.

"	...probably there is no word of the kind which has undergone such extraordinary variety of modifications, whilst retaining the same meaning, as this...	"
— Hobson-Jobson (1886)		

In the eastern Mediterranean, bādinjān was borrowed into Byzantine Greek first as ματιζάνιον, then modified to μελιτζάνα melitzána and melanzana, influenced by Greek μελανο- ‹black›. This came into Italian as melongiana and melanzana, and into Medieval Latin as melongena. The Latin name was later used by Tournefort as a genus name, then by Linnaeus as a species name. These forms came into English; though melongene has become obsolete, as have the French merangène, melongène/melanjan, it persists in the Caribbean English melongene or meloongen. In Italian, melanzana was interpreted as mela insana 'crazy apple'; this was translated into English as mad apple.

Cultivated Varieties

The plant and fruit have a profusion of English names: eggplant (North America, Australia), aubergine (Britain), brinjal (South Asia, South Africa, Malaysia, Singapore, terong (Indonesia), and West Indies, Trinidad), melongene (Caribbean), and formerly melongena and mad-apple.

Three varieties of eggplant

Different varieties of the plant produce fruit of different size, shape, and color, though typically purple. The most widely cultivated varieties—cultivars—in Europe and North America today are elongated ovoid, 12–25 cm long (4 $\frac{1}{2}$ to 9 in) and 6–9 cm broad (2 to 4 in) with a dark purple skin.

A much wider range of shapes, sizes and colors is grown in India and elsewhere in Asia. Larger varieties weighing up to a kilogram (2.2 pounds) grow in the region between the Ganges and Yamuna rivers, while smaller varieties are found elsewhere. Colors vary from white to yellow or green, as well as reddish-purple and dark purple. Some cultivars have a color gradient—white at the stem; to bright pink, deep purple or even black. Green or purple cultivars with white striping also exist. Chinese varieties are commonly shaped like a narrower, slightly pendulous cucumber, and are sometimes miscalled Japanese aubergines in North America. But there are also Asian varieties of Japanese breeding.

- Oval or elongated oval-shaped and black-skinned cultivars include 'Harris Special Hibush', 'Burpee Hybrid', 'Bringal Bloom', 'Black Magic', 'Classic', 'Dusky', and 'Black Beauty'.
- Slim cultivars in purple-black skin include 'Little Fingers', 'Ichiban', 'Pingtung Long', and 'Tycoon'
    - In green skin, 'Louisiana Long Green' and 'Thai (Long) Green'
    - In white skin, 'Dourga'.
- Traditional, white-skinned, egg-shaped cultivars include 'Casper' and 'Easter Egg'.
- Bi-colored cultivars with color gradient include 'Rosa Bianca', 'Violetta di Firenze', 'Bianca Sfumata di Rosa' (heirloom), and 'Prosperosa' (heirloom).
- Bi-colored cultivars with striping include 'Listada de Gandia' and 'Udumalapet'.
- In some parts of India, miniature varieties, most commonly called vengan, are popular.

Varieties

- Solanum melongena var. esculentum common aubergine, including white varieties, with many cultivars
- Solanum melongena var. depressum dwarf aubergine
- Solanum melongena var. serpentium snake aubergine

Genetically Engineered Aubergine

Bt brinjal is a transgenic aubergine that contains a gene from the soil bacterium Bacillus thuringiensis. This variety was designed to give the plant resistance to lepidopteran insects like the brinjal fruit and shoot borer (Leucinodes orbonalis) and fruit borer (Helicoverpa armigera).

On 9 February 2010, the Environment Ministry of India imposed a moratorium on the cultivation of Bt brinjal after protests against regulatory approval of cultivated Bt brinjal in 2009, stating the moratorium would last "for as long as it is needed to establish public trust and confidence".This decision was deemed controversial, as it deviated from previous practices with other genetically modified crops in India. Bt brinjal was approved for commercial cultivation in Bangaladesh in 2013.

Cultivation and Pests

In tropical and subtropical climates, eggplant can be sown directly into the garden. Eggplant grown in temperate climates fares better when transplanted into the garden after all danger of frost has passed. Seeds are typically started eight to ten weeks prior to the anticipated frost-free date. Solanum melongena is included on a list of low flammability plants, indicating that it is suitable for growing within a building protection zone.

Spacing should be 45 to 60 cm (18 to 24 in) between plants, depending on cultivar, and 60 to 90 cm (24 to 35 in) between rows, depending on the type of cultivation equipment being used. Mulching helps conserve moisture and prevent weeds and fungal diseases. The flowers are relatively unattractive to bees and the first blossoms often do not set fruit. Hand pollination by shaking the flowers improves the set of the first blossoms. Growers typically cut fruits from the vine just above the calyx owing to the somewhat woody stems. Flowers are complete, containing both female and male structures, and may be self-pollinated or cross-pollinated.

Many of the pests and diseases that afflict other Solanaceous plants, such as tomato, pepper (capsicum), and potato, are also troublesome to eggplants. For this reason, it should generally not be planted in areas previously occupied by its close relatives. However, since eggplants can be particularly susceptible to pests such as white flies, they are sometimes grown with slightly less susceptible plants, such as pepper, as a sacrificial trap crop. Four years should separate successive crops of eggplants to reduce pest pressure.

Common North American pests include the potato beetles, flea beetles, aphids, white flies and spider mites. Mature adult pests can be removed by hand, though flea beetles can be especially difficult to control. A commonly used herbicide for eggplant is dimethyl tetrachloroterephthalate. Good sanitation and crop rotation practices are extremely important for controlling fungal disease, the most serious of which is Verticillium.

Production

In 2013, global production of eggplants was 49.4 million tonnes. More than 1,600,000 hectares (4,000,000 acres) are devoted to the cultivation of eggplants in the world. 57% of output comes from China alone. India (27% of world total), Iran, Egypt and Turkey were also major producers that, combined with other Asian countries, constituted 94% of world production.

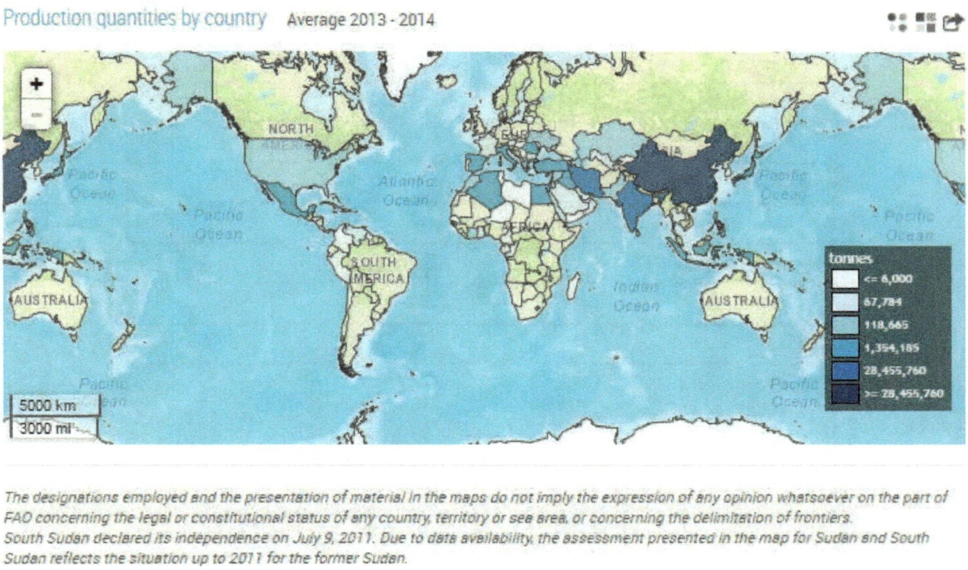

Production of eggplant in 2013 by country

Top countries in eggplant production (2013)(millions of tonnes)(million sq. mi)						
Rank **Country** **Production**			**Production/Area**			
1	China	28.4	7.68			
2	India	13.4	10.56			
3	Iran	1.4	2.2			
4	Egypt	1.2	3.08			
5	Turkey	0.8	2.64			
	World	49.4	0.85			

Nutrition

Eggplant, raw	
Nutritional value per 100 g (3.5 oz)	
Energy	104 kJ (25 kcal)

Carbohydrates	5.88 g
Sugars	3.53 g
Dietary fiber	3 g
Fat	0.18 g
Protein	0.98 g
Vitamins	
Thiamine (B1)	(3%) 0.039 mg
Riboflavin (B2)	(3%) 0.037 mg
Niacin (B3)	(4%) 0.649 mg
Pantothenic acid (B5)	(6%) 0.281 mg
Vitamin B6	(6%) 0.084 mg
Folate (B9)	(6%) 22 μg
Vitamin C	(3%) 2.2 mg
Vitamin E	(2%) 0.3 mg
Vitamin K	(3%) 3.5 μg
Minerals	
Calcium	(1%) 9 mg
Iron	(2%) 0.23 mg
Magnesium	(4%) 14 mg

Manganese	(11%) 0.232 mg
Phosphorus	(3%) 24 mg
Potassium	(5%) 229 mg
Zinc	(2%) 0.16 mg
Link to USDA Database entry	
• Units • μg = micrograms • mg = milligrams • IU = International units	
Percentages are roughly approximated using US recommendations for adults. Source: USDA Nutrient Database	

Nutritionally, raw eggplant is low in fat, protein, dietary fiber and carbohydrates (see table). It also provides low amounts of essential nutrients, with only manganese having a moderate percentage (11%) of the Daily Value (table). Minor changes in nutrient composition occur with season, environment of cultivation (open field or greenhouse), and genotype.

Chemistry

The color of purple skin varieties is due to the anthocyanin nasunin.

The browning of eggplant flesh results from the oxidation of polyphenols, such as the most abundant phenolic compound in the fruit, chlorogenic acid.

Allergies

Case reports of itchy skin or mouth, mild headache, and stomach upset after handling or eating eggplant have been reported anecdotally and published in medical journals (see also oral allergy syndrome). A 2008 study of a sample of 741 people in India, where eggplant is commonly consumed, found nearly 10% reported some allergic symptoms after consuming eggplant, with 1.4% showing symptoms within less than two hours. Contact dermatitis from eggplant leaves and allergy to eggplant flower pollen have also been reported. Individuals who are atopic (genetically predisposed to developing certain allergic hypersensitivity reactions) are more likely to have a reaction to eggplant, which may be because eggplant is high in histamines. A few proteins and at least one secondary metabolite have been identified as potential allergens. Cooking eggplant thoroughly seems to preclude reactions in some individuals, but at least one of the allergenic proteins survives the cooking process.

Taxonomy

The eggplant is quite often featured in the older scientific literature under the junior synonyms S. ovigerum and S. trongum. Several other now-invalid names have been uniquely applied to it:

- Melongena ovata Mill.
- Solanum album Noronha
- Solanum insanum L.
- Solanum longum Roxb.
- Solanum melanocarpum Dunal
- Solanum melongenum St.-Lag.
- Solanum oviferum Salisb.
- Prachi Salisb.

Segmented purple eggplant

A number of subspecies and varieties have been named, mainly by Dikii, Dunal, and (invalidly) by Sweet. Names for various eggplant types, such as agreste, album, divaricatum, esculentum, giganteum, globosi, inerme, insanum, leucoum, luteum, multifidum, oblongo-cylindricum, ovigera, racemiflorum, racemosum, ruber, rumphii, sinuatorepandum, stenoleucum, subrepandum, tongdongense, variegatum, violaceum and viride, are not considered to refer to anything more than cultivar groups at best. On the other hand, Solanum incanum and cockroach berry (S. capsicoides), other eggplant-like nightshades described by Linnaeus and Allioni, respectively, were occasionally considered eggplant varieties, but this is not correct.

Matured yellow eggplant

The eggplant has a long history of taxonomic confusion with the scarlet and Ethiopian eggplants (Solanum aethiopicum), known as gilo and nakati respectively, and described by Linnaeus as S. aethiopicum. The eggplant was sometimes considered a variety violaceum of that species. S. violaceum of de Candolle applies to Linnaeus' S. aethiopicum. There is an actual S. violaceum, an unrelated plant described by Ortega, which used to include Dunal's S. amblymerum and was often confused with the same author's S. brownii.

Like the potato and Solanum lichtensteinii, but unlike the tomato, which then was generally put in a different genus, the eggplant was also described as S. esculentum, in this case once more in the course of Dunal's work. He also recognized the varieties aculeatum, inerme and subinerme at that time. Similarly, H.C.F. Schuhmacher and Peter Thonning named the eggplant as S. edule, which is also a junior synonym of sticky nightshade (S. sisymbriifolium). Scopoli's S. zeylanicum refers to the eggplant, and that of Blanco to S. lasiocarpum.

BELL PEPPER

The bell pepper (also known as sweet pepper or pepper in the United Kingdom, Canada and Ireland, and capsicum /in Australia, India, Pakistan, Bangladesh, Singapore and New Zealand) is a cultivar group of the species Capsicum annuum. Cultivars of the plant produce fruits in different colors, including red, yellow, orange, green, chocolate/brown, vanilla/white, and purple. Bell peppers are sometimes grouped with less pungent pepper varieties as "sweet peppers". The ribs and seeds inside bell peppers may be consumed, but some people find the taste to be bitter.

Peppers are native to Mexico, Central America, and northern South America. Pepper seeds were imported to Spain in 1493, and from there spread to other European, African, and Asian countries. Today, China is the world's largest pepper producer, followed by Mexico and Indonesia.

Ideal growing conditions for bell peppers include warm soil, ideally 21 to 29 °C (70 to 84 °F), that is kept moist but not waterlogged. Bell peppers are sensitive to an abundance of moisture and extreme temperatures.

Red, yellow and green bell peppers

Nomenclature

The misleading name "pepper" was given by Europeans when Christopher Columbus brought the plant back to Europe. At that time, black pepper (peppercorns), from the unrelated plant Piper nigrum originating from India, was a highly prized condiment; the name "pepper" was at that time applied in Europe to all known spices with a hot and pungent taste and was therefore naturally extended to the newly discovered Capsicum genus. The most commonly used alternative name of the plant family, "chile", is of Mexican origin, from the Nahuatl word chilli. Botanically speaking, bell peppers are fruit, although they are considered vegetables in culinary contexts.

The bell pepper is the only member of the Capsicum genus that does not produce capsaicin, a lipophilic chemical that can cause a strong burning sensation when it comes in contact with mucous membranes. The lack of capsaicin in bell peppers is due to a recessive form of a gene that eliminates capsaicin and, consequently, the "hot" taste usually associated with the rest of the Capsicum genus. This recessive gene is overwritten in the Mexibelle pepper, a hybrid variety of bell pepper that produces small amounts of capsaicin (and is thus mildly pungent). Sweet pepper cultivars produce non pungent capsaicinoids, with many physiological effects similar to the more pungent sister compound capsaican.

The terms "bell pepper" (US), "pepper" (UK), and "capsicum" (India, Australia, and New Zealand) are often used for any of the large bell shaped fruits, regardless of their color. In British and Canadian English, the fruit is simply referred to as a "pepper", or additionally by color (as in the term "green pepper", for example), whereas in the United States and Malaysia, they are usually referred to as "bell peppers". In parts of the U.S. Midwest, bell peppers are called "mangoes." Canadian English uses both "bell pepper" and "pepper" interchangeably.

In some languages, the term "paprika", which has its roots in the word for pepper, is used for both the spice and the fruit – sometimes referred to by their color (e.g., "groene paprika", "gele paprika", in Dutch, which are green and yellow, respectively). The bell pepper is called papurika or "piiman, from Portuguese pimentão) in Japan. In Switzerland, the fruit is mostly called "peperone", which is the Italian name of the fruit. In France, it is called "poivron", with the same root as "poivre" (meaning "pepper") or "piment". In South Korea, the word " pimang from the Japanese "piiman) refers to green bell peppers, whereas papurika from paprika refers to bell peppers of other colors. In Sri Lanka, (only) the fruit used as a vegetable is called "maalu miris".

Colors

The most common colors of bell peppers are green, yellow, orange and red. More rarely, brown, white, lavender, and dark purple peppers can be seen, depending on the variety. Most typically, unripe fruits are green or, less commonly, pale yellow or purple. Red bell peppers are simply ripened green peppers,although the Permagreen variety maintains its green color even when fully ripe.

Peppers in five colors (Banana peppers are second from the left)

Green peppers are less sweet and slightly more bitter than yellow or orange peppers, with red bell peppers being the sweetest. The taste of ripe peppers can also vary with growing conditions and post-harvest storage treatment; the sweetest fruits are allowed to ripen fully on the plant in full sunshine, while fruit harvested green and after-ripened in storage is less sweet.

Nutritional Value

Peppers, sweet, green, raw	
Nutritional value per 100 g (3.5 oz)	
Energy	84 kJ (20 kcal)
Carbohydrates	4.64 g
Sugars	2.4 g
Dietary fiber	1.8 g
Fat	0.17 g
Protein	0.86 g
Vitamins	

Vitamin A equiv. beta-carotene lutein zeaxanthin	(2%) 18 μg (2%) 208 μg 341 μg
Thiamine (B1)	(5%) 0.057 mg
Riboflavin (B2)	(2%) 0.028 mg
Niacin (B3)	(3%) 0.48 mg
Pantothenic acid (B5)	(2%) 0.099 mg
Vitamin B6	(17%) 0.224 mg
Folate (B9)	(3%) 10 μg
Vitamin C	(97%) 80.4 mg
Vitamin E	(2%) 0.37 mg
Vitamin K	(7%) 7.4 μg
Minerals	
Calcium	(1%) 10 mg
Iron	(3%) 0.34 mg
Magnesium	(3%) 10 mg
Manganese	(6%) 0.122 mg

Phosphorus	(3%) 20 mg
Potassium	(4%) 175 mg
Sodium	(0%) 3 mg
Zinc	(1%) 0.13 mg
Other constituents	
Fluoride	2 µg
Link to USDA Database entry	
• Units • µg = micrograms • mg = milligrams • IU = International units	
Percentages are roughly approximated using US recommendations for adults. Source: USDA Nutrient Database	

Capsicum peppers are rich sources of antioxidants and vitamin C. The level of carotene, like lycopene, is nine times higher in red peppers. Red peppers have twice the vitamin C content of green peppers.

Red and green bell peppers are high in para-coumaric acid.

The characteristic aroma of green peppers is caused by 3-isoButyl-2-methoxypyrazine (IBMP). Its detection threshold in water is estimated to be 2 ng/L. The same chemical is responsible for characteristic Cabernet Sauvignon green note.

Production

Bell and Chili pepper production (metric tons)				
Country	**2004**	**2005**	**2006**	**2007**
China	12,031,031	12,530,180	13,031,000	14,033,000
Mexico	1,431,258	1,617,264	1,681,277	1,690,000
Indonesia	1,100,514	1,058,023	1,100,000	1,100,000
Turkey	1,700,000	1,829,000	1,842,175	1,090,921
Spain	1,077,025	1,063,501	1,074,100	1,065,000

United States	978,890	959,070	998,210	855,870
Nigeria	720,000	721,000	721,500	723,000
Egypt	467,433	460,000	470,000	475,000
South Korea	410,281	395,293	352,966	345,000
Netherlands	318,000	345,000	318,000	340,000
Romania	237,240	203,751	279,126	280,000
Ghana	270,000	270,000	277,000	279,000
Italy	362,430	362,994	345,152	252,194
Tunisia	255,000	256,000	256,000	250,000
Algeria	265,307	248,614	275,888	233,000
Hungary	126,133	113,371	206,419	207,000
Morocco	182,340	190,480	235,570	192,000
Serbia*	159,741	167,477	177,255	150,257
Japan	153,400	154,000	146,900	150,000
Israel	129,100	134,700	150,677	136,000
World	24,587,124	25,261,259	26,252,907	26,056,900

Note: Serbia before 2006 included Montenegro

POTATO

The potato is a starchy, tuberous crop from the perennial nightshade Solanum tuberosum. The word "potato" may refer either to the plant itself or to the edible tuber. In the Andes, where the species is indigenous, there are some other closely related cultivated potato species. Potatoes were introduced outside the Andes region approximately four centuries ago, and have since become an integral part of much of the world's food supply. It is the world's fourth-largest food crop, following maize, wheat, and rice.The green leaves and green skins of tubers exposed to the light are toxic.

Wild potato species can be found throughout the Americas from the United States to southern Chile. The potato was originally believed to have been domesticated independently in multiple locations, but later genetic testing of the wide variety of cultivars and wild species proved a single origin for potatoes in the area of present-day southern Peru and extreme northwestern Bolivia (from a species in the Solanum brevicaule complex), where they were domesticated approximately 7,000–10,000 years ago.Following centuries of selective breeding, there are now over a thousand different types of potatoes. Over 99% of the presently cultivated potatoes worldwide descended from varieties that originated in the lowlands of south-central Chile, which have displaced formerly popular varieties from the Andes.

However, the local importance of the potato is variable and changing rapidly. It remains an essential crop in Europe (especially eastern and central Europe), where per capita production is still the highest in the world, but the most rapid expansion over the past few decades has occurred in southern and eastern Asia. As of 2007 China led the world in potato production, and nearly a third of the world's potatoes were harvested in China and India.

Potato cultivars appear in a variety of colors, shapes, and sizes

Characteristics

Potato plants are herbaceous perennials that grow about 60 cm (24 in) high, depending on variety, with the leaves dying back after flowering, fruiting and tuber formation. They bear white, pink, red, blue, or purple flowers with yellow stamens. In general, the tubers of varieties with white flowers have white skins, while those of varieties with colored flowers tend to have pinkish skins.

Flowers of a potato plant

Russet potatoes

Potatoes are mostly cross-pollinated by insects such as bumblebees, which carry pollen from other potato plants, though a substantial amount of self-fertilizing occurs as well. Tubers form in response to decreasing day length, although this tendency has been minimized in commercial varieties.

Potato plants

After flowering, potato plants produce small green fruits that resemble green cherry tomatoes, each containing about 300 seeds. Like all parts of the plant except the tubers, the fruit contain the toxic alkaloidsolanine and are therefore unsuitable for consumption. All new potato varieties are grown from seeds, also called “true potato seed”, “TPS” or “botanical seed” to distinguish it from seed tubers. New varieties grown from seed can be propagated vegetatively by planting tubers, pieces of tubers cut to include at least one or two eyes, or cuttings, a practice used in greenhouses for the production of healthy seed tubers. Plants propagated from tubers are clones of the parent, whereas those propagated from seed produce a range of different varieties.

Genetics

There are about 5,000 potato varieties worldwide. Three thousand of them are found in the Andes alone, mainly in Peru, Bolivia, Ecuador, Chile, and Colombia. They belong to eight or nine species, depending on the taxonomic school. Apart from the 5,000 cultivated varieties, there are about 200 wild species and subspecies, many of which can be cross-bred with cultivated varieties. Cross-breeding has been done repeatedly to transfer resistances to certain pests and diseases from the gene pool of wild species to the gene pool of cultivated potato species. Genetically modified varieties have met public resistance in the United States and in the European Union.

The major species grown worldwide is Solanum tuberosum (a tetraploid with 48 chromosomes), and modern varieties of this species are the most widely cultivated. There are also four diploid species (with 24 chromosomes): S. stenotomum, S. phureja, S. goniocalyx, and S. ajanhuiri. There are two triploid species (with 36 chromosomes): S. chaucha and S. juzepczukii. There is one pentaploid cultivated species (with 60 chromosomes): S. curtilobum. There are two major subspecies of Solanum tuberosum: andigena, or Andean; and tuberosum, or Chilean. The Andean potato is adapted to the short-day conditions prevalent in the mountainous equatorial and tropical regions where it originated; the Chilean potato, however, native to the Chiloé Archipelago, is adapted to the long-day conditions prevalent in the higher latitude region of southern Chile.

The International Potato Center, based in Lima, Peru, holds an ISO-accredited collection of potato germplasm. The international Potato Genome Sequencing Consortium announced in 2009 that they had achieved a draft sequence of the potato genome. The potato genome contains 12 chromosomes and 860 million base pairs, making it a medium-sized plant genome. More than 99 percent of all current varieties of potatoes currently grown are direct descendants of a subspecies that once grew in the lowlands of south-central Chile. Nonetheless, genetic testing of the wide variety of cultivars and wild species affirms that all potato subspecies derive from a single origin in the area of present-day southern Peru and extreme northwestern Bolivia (from a species in the Solanum brevicaule complex).

Most modern potatoes grown in North America arrived through European settlement and not independently from the South American sources; however, at least one wild potato species, Solanum fendleri, is found as far north as Texas and is used in breeding for resistance to a nematode species that attacks cultivated potatoes. A secondary center of genetic variability of the potato is Mexico, where important wild species that have been used extensively in modern breeding are found, such as the hexaploid Solanum demissum, as a source of resistance to the devastating late blight disease. Another relative native to this region, Solanum bulbocastanum, has been used to genetically engineer the potato to resist potato blight.

Potatoes yield abundantly with little effort, and adapt readily to diverse climates as long as the climate is cool and moist enough for the plants to gather sufficient water from the soil to form the starchy tubers. Potatoes do not keep very well in storage and are vulnerable to molds that feed on the stored tubers and quickly turn them rotten, however: crops such as grain can be stored for several years with a low risk of rot. The yield of Calories per acre (about 9.2 million) is higher than that of maize (7.5 million), rice (7.4 million), wheat (3 million), or soybean (2.8 million).

Growth and Cultivation

Potatoes are generally grown from seed potatoes – these are tubers specifically grown to be disease free and provide consistent and healthy plants. To be disease free, the areas where seed potatoes are grown are selected with care. In the USA this restricts production of seed potatoes to only 15 states out of the 50 states that grow potatoes. These locations are selected for their cold hard winters that kill pests and long sunshine hours in the summer for optimum growth. In the UK, most seed potatoes originate in Scotland in areas where westerly winds prevent aphid attack and thus prevent spread of potato virus pathogens.Potato growth has been divided into five phases.

Potato planting

Potato field in Fort Fairfield, Maine

Potatoes grown in a tall bag are common in gardens as they minimize the amount of digging required at harvest

During the first phase, sprouts emerge from the seed potatoes and root growth begins. During the second, photosynthesis begins as the plant develops leaves and branches. In the third phase stolons develop from lower leaf axils on the stem and grow downwards into the ground and on these stolons new tubers develop as swellings of the stolon. This phase is often (but not always) associated with flowering. Tuber formation halts when soil temperatures reach 27 °C (81 °F); hence potatoes are considered a cool-season crop. Tuber bulking occurs during the fourth phase, when the plant begins investing the majority of its resources in its newly formed tubers. At this stage, several factors are critical to yield: optimal soil moisture and temperature, soil nutrient availability and balance, and resistance to pest attacks. The final phase is maturation: The plant canopy dies back, the tuber skins harden, and their sugars convert to starches.

New tubers may arise at the soil surface. Since exposure to light leads to greening of the skins and the development of solanine, growers are interested in covering such tubers. Commercial growers usually address this problem by piling additional soil around the base of the plant as it grows (hilling, or in British English earthing up). An alternative method used by home gardeners and smaller-scale growers involves covering the growing area with organic mulches such as straw or with plastic sheets.

Correct potato husbandry can be an arduous task in some circumstances. Good ground preparation, harrowing, plowing, and rolling are always needed, along with a little grace from the weather and a good source of water. Three successive plowings, with associated harrowing and rolling, are desirable before planting. Eliminating all root-weeds is desirable in potato cultivation. In general, the potatoes themselves are grown from the eyes of another potato and not from seed. Home gardeners often plant a piece of potato with two or three eyes in a hill of mounded soil. Commercial growers plant potatoes as a row crop using seed tubers, young plants or microtubers and may mound the entire row. Seed potato crops are ‹rogued› in some countries to eliminate diseased plants or those of a different variety from the seed crop.

Potatoes are sensitive to heavy frosts, which damage them in the ground. Even cold weather makes potatoes more susceptible to bruising and possibly later rotting, which can quickly ruin a large stored crop.

At harvest time, gardeners usually dig up potatoes with a long-handled, three-prong grape (or graip), i.e., a spading fork, or a potato hook, which is similar to the graip but with tines at a 90° angle to the handle. In larger plots, the plow is the fastest implement for unearthing potatoes. Commercial harvesting is typically done with large potato harvesters, which scoop up the plant and surrounding earth. This is transported up an apron chain consisting of steel links several feet wide, which separates some of the dirt. The chain deposits into an area where further separation occurs. Different designs use different systems at this point. The most complex designs use vine choppers and shakers, along with a blower system to separate the potatoes from the plant. The result is then usually run past workers who continue to sort out plant material, stones, and rotten potatoes before the potatoes are continuously delivered to a wagon or truck. Further inspection and separation occurs when the potatoes are unloaded from the field vehicles and put into storage.

Immature potatoes may be sold as new potatoes and are particularly valued for taste. These are often harvested by the home gardener or farmer by grabbling, i.e. pulling out the young tubers by hand while leaving the plant in place.

Potatoes are usually cured after harvest to improve skin-set. Skin-set is the process by which the skin of the potato becomes resistant to skinning damage. Potato tubers may be susceptible to skinning at harvest and suffer skinning damage during harvest and handling operations. Curing allows the skin to fully set and any wounds to heal. Wound-healing prevents infection and water-loss from the tubers during storage. Curing is normally done at relatively warm temperatures 50 to 60 °F (10 to 16 °C) with high humidity and good gas-exchange if at all possible.

Potato plant prior to harevest.

Storage

Storage facilities need to be carefully designed to keep the potatoes alive and slow the natural process of decomposition, which involves the breakdown of starch. It is crucial that the storage area is dark, well ventilated and for long-term storage maintained at temperatures near 4 °C (39 °F). For short-term storage before cooking, temperatures of about 7 to 10 °C (45 to 50 °F) are preferred.

On the other hand, temperatures below 4 °C (39 °F) convert potatoes› starch into sugar, which alters their taste and cooking qualities and leads to higher acrylamide levels in the cooked product, especially in deep-fried dishes—the discovery of acrylamides in starchy foods in 2002 has led to many international health concerns as they are believed to be possible carcinogens and their occurrence in cooked foods is currently under study as a possible influence in potential health problems.[a]

Under optimum conditions possible in commercial warehouses, potatoes can be stored for up to ten to twelve months. When stored in homes, the shelf life is usually only a few weeks. If potatoes develop green areas or start to sprout, these areas should be trimmed before using.Trimming or peeling green areas are inadequate to remove copresent toxins, and such potatoes are no longer suitable as animal food.

Commercial storage of potatoes involves several phases: drying of surface moisture; a wound healing phase at 85% to 95% relative humidity and temperatures below 25 °C (77 °F); a staged cooling phase; a holding phase; and a reconditioning phase, during which the tubers are slowly warmed. Mechanical ventilation is used at various points during the process to prevent condensation and accumulation of carbon dioxide.

Yield

The world dedicated 18.6 million hectares in 2010 for potato cultivation. The average world farm yield for potato was 17.4 tonnes per hectare, in 2010. Potato farms in the United States were the most productive in 2010, with a nationwide average of 44.3 tonnes per hectare. United Kingdom was a close second.

New Zealand farmers have demonstrated some of the best commercial yields in the world, ranging between 60 and 80 tonnes per hectare, some reporting yields of 88 tonnes potatoes per hectare.

There is a big gap among various countries between high and low yields, even with the same variety of potato. Average potato yields in developed economies ranges between 38–44 tonnes per hectare. China and India accounted for over a third of world's production in 2010, and had yields of 14.7 and 19.9 tonnes per hectare respectively. The yield gap between farms in developing economies and developed economies represents an opportunity loss of over 400 million tonnes of potato, or an amount greater than 2010 world potato production. Potato crop yields are determined by factors such as the crop breed, seed age and quality, crop management practices and the plant environment. Improvements in one or more of these yield determinants, and a closure of the yield gap, can be a major boost to food supply and farmer incomes in the developing world.

Varieties

While there are close to 4,000 varieties of potato, it has been bred into many standard or well-known varieties, each of which has particular agricultural or culinary attributes. In general, varieties are categorized into a few main groups, such as russets, reds, whites, yellows (also called Yukons) and purples—based on common characteristics. Around 80 varieties are commercially available in the UK.For culinary purposes, varieties are often differentiated by their waxiness. Floury, or mealy (baking) potatoes have more starch (20–22%) than waxy (boiling) potatoes (16–18%). The distinction may also arise from variation in the comparative ratio of two potato starch compounds: amylose and amylopectin. Amylose, a long-chain molecule, diffuses from the starch granule when cooked in water, and lends itself to dishes where the potato is mashed. Varieties that contain a slightly higher amylopectin content, a highly branched molecule, help the potato retain its shape when boiled.

Bamberg potatoes

Organically grown Russet Burbanks

The European Cultivated Potato Database (ECPD) is an online collaborative database of potato variety descriptions, updated and maintained by the Scottish Agricultural Science Agency within the framework of the European Cooperative Programme for Crop Genetic Resources Networks (ECP/GR)—which is run by the International Plant Genetic Resources Institute (IPGRI).

Popular varieties (cultivars) include:

• Adirondack Blue	• Clavela Blanca	• Norgold Russet
• Adirondack Red	• Désirée	• Pachacoña
• Agata	• Estima	• Pike
• Almond	• Fianna	• Pink Eye
• Alpine Russet	• Fingerling	• Pink Fir Apple
• Alturas	• Flava	• Primura
• Amandine	• French Fingerling	• Ranger Russet
• Annabelle	• German Butterball	• Ratte
• Anya	• Golden Wonder	• Record
• Arran Victory	• Goldrush	• Red La Soda
• Atlantic	• Home Guard	• Red Norland
• Austrian Crescent	• Innovator	• Red Pontiac
• Avalanche	• Irish Cobbler	• Rooster
• Bamberg	• Irish Lumper	• Russet Burbank
• Bannock Russet	• Jersey Royal	• Russet Norkotah
• Belle de Fontenay	• Kennebec	• Selma
• BF-15	• Kerr's Pink	• Shepody
• Bildtstar	• Kestrel	• Sieglinde
• Bintje	• Keuka Gold	• Silverton Russet
• Blazer Russet	• King Edward	• Sirco
• Blue Congo	• Kipfler	• Snowden
• Bonnotte	• Lady Balfour	• Spunta
• British Queen	• Langlade	• Up to date
• Cabritas	• Linda potato	• Stobrawa
• Camota	• Marcy	• Superior
• Canela Russet	• Marfona	• Umatilla Russet
• Cara	• Maris Piper	• Villetta Rose
• Carola	• Marquis	• Vivaldi
• Chelina	• Megachip	• Vitelotte
• Chiloé	• Melody	• Yellow Finn
• Cielo	• Monalisa	• Yukon Gold
	• Nicola	

Blue Varieties

The blue or purple potato originated in South America. It has purple skin and flesh, which becomes blue once cooked. It has a slight whitish scab that seems to be present in all samples. The variety, called "Cream of the Crop", has been introduced into Ireland and has proved popular. To preserve the purple color best, microwaving is the best way to cook it; however, it can also be baked and steamed. The purple or blue potato tends to have a more nutty flavor than its relatives of other colors.

Potato variety "Blue Swede"

A mutation in the varieties' P locus causes production of the antioxidant anthocyanin.

Genetically Engineered Potatoes

Genetic research has produced several genetically modified varieties. 'New Leaf', owned by Monsanto Company, incorporates genes from Bacillus thuringiensis, which confers resistance to the Colorado potato beetle; 'New Leaf Plus' and 'New Leaf Y', approved by US regulatory agencies during the 1990s, also include resistance to viruses. McDonald's, Burger King, Frito-Lay, and Procter & Gamble announced they would not use genetically modified potatoes, and Monsanto published its intent to discontinue the line in March 2001.

Waxy potato varieties produce two main kinds of potato starch, amylose and amylopectin, the latter of which is most industrially useful. The German chemical company BASF created the Amflora potato, which has been modified to contain antisense against the enzyme that drives synthesis of amylose, namely granule bound starch synthase. This resulting potato almost exclusively produces amylopectin, and thus is more useful for the starch industry. In 2010, the European Commission cleared the way for 'Amflora' to be grown in the European Union for industrial purposes only—not for food. Nevertheless, under EU rules, individual countries have the right to decide whether they will allow this potato to be grown on their territory. Commercial planting of 'Amflora' was expected in the Czech Republic and Germany in the spring of 2010, and Sweden and the Netherlands in subsequent years. Another GM potato variety developed by BASF is 'Fortuna' which was made resistant to late blight by adding two resistance genes, blb1 and blb2, which originate from the Mexican wild potato Solanum bulbocastanum. In October 2011 BASF requested cultivation and marketing approval as a feed and food from the EFSA. In 2012, GMO development in Europe was stopped by BASF.

In November 2014, the USDA approved a genetically modified potato developed by J.R. Simplot Company, which contains genetic modifications that prevent bruising and produce less acrylamide when fried than conventional potatoes; the modifications do not cause new proteins to be made, but rather prevent proteins from being made via RNA interference.

Pests

The historically significant Phytophthora infestans (late blight) remains an ongoing problem in Europeand the United States. Other potato diseases include Rhizoctonia, Sclerotinia, black leg, powdery mildew, powdery scab and leafroll virus.

A potato ruined by late blight

Insects that commonly transmit potato diseases or damage the plants include the Colorado potato beetle, the potato tuber moth, the green peach aphid (Myzus persicae), the potato aphid, beet leafhoppers, thrips, and mites. The potato cyst nematode is a microscopic worm that thrives on the roots, thus causing the potato plants to wilt. Since its eggs can survive in the soil for several years, crop rotation is recommended.

Pesticides

During the crop year 2008, many of the certified organic potatoes produced in the United Kingdom and certified by the Soil Association as organic were sprayed with a copper pesticide to control potato blight (Phytophthora infestans). According to the Soil Association, the total copper that can be applied to organic land is 6 kg/ha/year.

According to an Environmental Working Group analysis of USDA and FDA pesticide residue tests performed from 2000 through 2008, 84% of the 2,216 tested potato samples contained detectable traces of at least one pesticide. A total of 36 unique pesticides were detected on potatoes over the 2,216 samples, though no individual sample contained more than 6 unique pesticide traces, and the average was 1.29 detectable unique pesticide traces per sample. The average quantity of all pesticide traces found in the 2,216 samples was 1.602 ppm. While this was a very low value of pesticide residue, it was the highest amongst the 50 vegetables analyzed.

Uses

- Potatoes are used to brew alcoholic beverages such as vodka, potcheen, or akvavit.
- They are also used as food for domestic animals.
- Potato starch is used in the food industry as, for example, thickeners and binders of soups and sauces, in the textile industry, as adhesives, and for the manufacturing of papers and boards.
- Maine companies are exploring the possibilities of using waste potatoes to obtain polylactic acid for use in plastic products; other research projects seek ways to use the starch as a base for biodegradable packaging.
- Potato skins, along with honey, are a folk remedy for burns in India. Burn centers in India have experimented with the use of the thin outer skin layer to protect burns while healing.
- Potatoes (mainly Russets) are commonly used in plant research. The consistent parenchyma tissue, the clonal nature of the plant and the low metabolic activity provide a very nice "model tissue" for experimentation. Wound-response studies are often done on potato tuber tissue, as are electron transport experiments. In this respect, potato tuber tissue is similar to Drosophila melanogaster, Caenorhabditis elegans and Escherichia coli: they are all "standard" research organisms.
- Potatoes have been delivered with personalized messages as a novelty. Potato delivery services include Potato Parcel and Mail A Spud.

REFERENCES

1. "Solanaceae Juss., nom. cons.". Germplasm Resources Information Network. United States Department of Agriculture. 2007-04-12. Retrieved 2009-04-16.
2. Olmstead, R. G.; Sweere, J. A.; Spangler, R. E.; Bohs, L.; Palmer, J. D. (1999). "Phylogeny and provisional classification of the Solanaceae based on chloroplast DNA" (PDF). In Nee, M.; Symon, D. E.; Lester, R. N.; Jessop, J. P. Solanaceae

IV: advances in biology and utilization. The Royal Botanic Gardens. pp. 111–37.

3. Olmstead, R.G.; Bohs, L. (2007). "A Summary of molecular systematic research in Solanaceae: 1982-2006". Acta Horticulturae. 745: 255–68.
4. Merriam-Webster, Merriam-Webster's Unabridged Dictionary, Merriam-Webster.
5. Yasin J. Nasir. "Solanaceae". Flora of Pakistan.
6. Fujii, Kenjiro (1934). Cytologia. Botanical Institute. p. 281.
7. Hunziker, A.T. 1979: South American Solanaceae: a synoptic review. In: D'ARCY, W.G., 1979: The Biology and Taxonomy of the Solanaceae. Linn. Soc. Symp. Ser. 7: p 48-85. Linnean Soc. & Academic Press; London.
8. Balken, J.A. THE PLANT FAMILY SOLANACEAE: FRUITS IN SOLANACEAE
9. Zeiger, E. 1998. Solanine and Chaconine. Review of Toxicological Literature. Integrated Laboratory Systems, USA.
10. "Solanine poisoning". Br Med J. 2 (6203): 1458–9. 1979-12-08. doi:10.1136/bmj.2.6203.1458-a. PMC 1597169Freely accessible. PMID 526812.
11. Alexander RF, Forbes GB, Hawkins ES (1948-09-11). "A Fatal Case of Solanine Poisoning". Br Med J. 2 (4575): 518. doi:10.1136/bmj.2.4575.518. PMC 2091497Freely accessible. PMID 18881287.
12. Griffin WJ, Lin GD (March 2000). "Chemotaxonomy and geographical distribution of tropane alkaloids". Phytochemistry. 53 (6): 623–37. doi:10.1016/S0031-9422(99)00475-6. PMID 10746874.
13. Sneden, A. The tropane alkaloids. Medicinal Chemistry and Drug Design. Virginia Commonwealth University [dead link]
14. Evans, W.C. 1979. Tropane alkaloids of the Solanaceae. En: HAWKES, LESTER and SHELDING (eds.). The biology and taxonomy of the Solanaceae. Linn. Soc. Symp. Ser. 7:241-254. Linnean Soc. & Academic Press., London.
15. Matsuda, Jun; Okabe, Souichi; Hashimoto, Takashi; Yamada, Yasuyuki (1991). "Molecular cloning of hyoscyamine 6β-hydroxylase, a 2-oxoglutarate-dependent dioxygenase, from cultured roots of Hyoscyamus niger". The Journal of Biological Chemistry. 266 (15): 9460–4. PMID 2033047.
16. Rocha, Pedro; Stenzel, Olaf; Parr, Adrian; Walton, Nicholas; Christou, Paul; Dräger, Birgit; Leech, Mark J (June 2002). "Functional expression of tropinone reductase I (trI) and hyoscyamine-6β-hydroxylase (h6h) from Hyoscyamus niger in Nicotiana tabacum". Plant Science. 162 (6): 905–13. doi:10.1016/S0168-9452(02)00033-X.
17. Cardillo, Alejandra B.; Giulietti, Ana M.; Marconi, Patricia L. (June 2006). "Analysis and sequencing of h6hmRNA, last enzyme in the tropane alkaloids pathway from anthers and hairy root cultures of Brugmansia candida (Solanaceae)". Electronic Journal of Biotechnology. 9 (3). doi:10.4067/S0717-34582006000300004.
18. Olmstead, Richard G.; Palmer, Jeffrey D. (1992). "A Chloroplast DNA Phylogeny of the Solanaceae: Subfamilial Relationships and Character Evolution". Annals

of the Missouri Botanical Garden. 79 (2): 346–60. doi:10.2307/2399773. JSTOR 2399773.

19. Martins, Talline R.; Barkman, Todd J. (2005). "Reconstruction of Solanaceae Phylogeny Using the Nuclear Gene SAMT". Systematic Botany. 30 (2): 435–47. doi:10.1600/0363644054223675. JSTOR 25064071.
20. Olmstead, R.G.; Bohs, L. (2007). "A Summary of molecular systematic research in Solanaceae: 1982-2006". Acta Horticulturae. 745: 255–68.
21. Ando, Toshio; Kokubun, Hisashi; Watanabe, Hitoshi; Tanaka, Norio; Yukawa, Tomohisa; Hashimoto, Goro; Marchesi, Eduardo; Suárez, Enrique; Basualdo, Isabel L. (2005). "Phylogenetic Analysis of Petunia sensu Jussieu (Solanaceae) using Chloroplast DNA RFLP". Annals of Botany. 96 (2): 289–97. doi:10.1093/aob/mci177. PMC 4246877Freely accessible. PMID 15944177.
22. Mishiba, Kei-Ichiro; Ando, Toshio; Mii, Masahiro; Watanabe, Hitoshi; Kokubun, Hisashi; Hashimoto, Goro; Marchesi, Eduardo (2000). "Nuclear DNA Content as an Index Character Discriminating Taxa in the Genus Petunia sensu Jussieu (Solanaceae)". Annals of Botany. 85 (5): 665–73. doi:10.1006/anbo.2000.1122.
23. Pérez, Fernanda; Arroyo, Mary T. K.; Medel, Rodrigo; Hershkovitz, Mark A. (2006). "Ancestral reconstruction of flower morphology and pollination systems in Schizanthus (Solanaceae)". American Journal of Botany. 93 (7): 1029–38. doi:10.3732/ajb.93.7.1029. PMID 21642168.
24. Garcia, Vicente F.; Olmstead, Richard G. (June 2003). "Phylogenetics of Tribe Anthocercideae (Solanaceae) Based on ndhF and trnL/F Sequence Data". Systematic Botany. 28 (3): 609–15. doi:10.1043/02-52.1 (inactive 2015-01-27). JSTOR 25063900.
25. "The Plant List, Atropa". Royal Botanic Garden, Kew.
26. Mace, E. S.; Gebhardt, C. G.; Lester, R. N. (1999). "AFLP analysis of genetic relationships in the tribe Datureae (Solanaceae)". Theoretical and Applied Genetics. 99 (3–4): 634–41. doi:10.1007/s001220051278. PMID 22665199.
27. Knapp, Sandra; Persson, Viveca; Blackmore, Stephen (1997). "A Phylogenetic Conspectus of the Tribe Juanulloeae (Solanaceae)". Annals of the Missouri Botanical Garden. 84 (1): 67–89. doi:10.2307/2399954. JSTOR 2399954.
28. Sazima, M.; Buzato, S; Sazima, I (2003). "Dyssochroma viridiflorum (Solanaceae): A Reproductively Bat-dependent Epiphyte from the Atlantic Rainforest in Brazil". Annals of Botany. 92 (5): 725–30. doi:10.1093/aob/mcg190. PMC 4244854Freely accessible. PMID 14500325.
29. Bernardello, Luis M. (1987). "Comparative Floral Morphology in Lycieae (Solanaceae)". Brittonia. 39 (1): 112–29. doi:10.2307/2806983. JSTOR 2806983.
30. Levin, achel A.; Mille, Jill S. (2005). "Relationships within tribe Lycieae (Solanaceae): Paraphyly of Lycium and multiple origins of gender dimorphism". American Journal of Botany. 92 (12): 2044–53. doi:10.3732/ajb.92.12.2044. JSTOR 4125537. PMID 21646122.
31. Bernardello, L.; Chiang-Cabrera, F. (1998). "A cladistic study on the American

species of Lycium (Solanaceae) based on morphological variation". In Fortunato, Renée H; Bacigalupo, Nélida M. Proceedings of the VI Congreso Latinoamericano de Botánica, Mar del Plata, Argentina, 2-8 October, 1994. Monographs in Systematic Botany from the Missouri Botanical Garden. Missouri Botanical Garden Press. pp. 33–46. ISBN 978-0-915279-58-6.

32. Smith, Stacey DeWitt; Baum, David A. (2006). "Phylogenetics of the florally diverse Andean clade Iochrominae (Solanaceae)". American Journal of Botany. 93 (8): 1140–53. doi:10.3732/ajb.93.8.1140. JSTOR 4122802. PMID 21642180.
33. Whitson, Maggie; Manos, Paul S. (2005). "Untangling Physalis (Solanaceae) from the Physaloids: A Two-Gene Phylogeny of the Physalinae". Systematic Botany. 30 (1): 216–30. doi:10.1600/0363644053661841. JSTOR 25064051.
34. Masis, C. & Madrigal, R. 1994. Lista preliminar de malezas hospedantes de Thrips (Thysanoptera) que dañan al Chrysanthemum morifolium en el valle central de Costa Rica. Agronomía Costarricense 18(1): 99-101. 1994
35. Ormeño, J., Sepúlveda R., Rojas, R. Malezas del género Datura como factor epidemiológico del virus del mosaico de la alfalfa (amv), virus del mosaico del pepino (cmv) y virus y de la papa (pvy) en Solanáceas cultivadas. Agricultura técnica Vol. 66, N°. 4, 2006, 333-341 Summary in Spanish
36. Pedrosa-Macedo, J., Olckers, T. & Vitorino, M. 2003. Phytophagous arthropods associated with Solanum mauritianum Scopoli (Solanaceae) in the first Plateau of Paraná, Brazil: a cooperative project on biological control of weeds between Brazil and South Africa. Neotrop. Entomol. 32: 519-522. Article in English, with a summary in Portuguese
37. Arboles ornamentales cultivados en España. Solanáceas
38. Shaw, J. 2007. A new hybrid genus for Calibrachoa × Petunia (Solanaceae). HANBURYANA 2: 50–51
39. The Value of Growing Petchoa SuperCal®. Ornamental News Oct 25 2012
40. International Tomato Sequencing Project Home
41. International Solanaceae Genomics Project (SOL), Systems Approach to Diversity and Adaptation.

CHAPTER 3

Brassicaceae Farming

Chapter Contents

Formerly known as cruciferae, brassicaceae is a family of plants most commonly referred to as the mustard or the cabbage family. A large and important family, brassicaceae contains many common vegetables, such as cabbage and broccoli, as well as ornamental plants, such as alyssum, and weeds, like bittercress. The mustard family consists of approximately 330 genera and 3,700 species.

Brassicaceae plants can usually be easily identified by their flowers. Species generally have clusters of flowers with four petals forming a cross shape. Flowers are usually white or yellow but can be red, blue, orange, or pink. Ornamental flowers, particularly, have a greater variety in their petal color. The cruciform arrangement of petals is what gave this family its original name, cruciferae.

Annual or perennial, brassicaceae plants usually have two chambered seed pods and alternating leaves climbing up their stems. These plants are found mostly in temperate regions in both the northern and the southern hemispheres. With a generally worldwide distribution, brassicaceae is found most diversely in the Mediterranean region. Many species also possess glycosinolate, which gives them the odor distinctive to broccoli or cabbage.

INTRODUCTION

Brassicaceae or Cruciferae is a medium-sized and economically important family of flowering plants commonly known as the mustards, the crucifers, or the cabbage family. The name *Brassicaceae* is derived from the included genus *Brassica*. The alternative older name, Cruciferae (English /kru:`srfəri/), meaning «cross-bearing», describes the four petals of mustard flowers, which resemble a cross. Cruciferae is one of eight plant family names without the suffix *-aceae* that are authorized alternative names (according to ICBN Art. 18.5 and 18.6 Vienna Code).

The family contains 372 genera and 4060 accepted species. The largest genera are *Draba* (440 species), *Erysimum* (261 species), *Lepidium* (234 species), *Cardamine* (233 species), and *Alyssum* (207 species).

The family contains the cruciferous vegetables, including species such as *Brassica oleracea* (e.g., broccoli, cabbage, cauliflower, kale, collards), *Brassica rapa* (turnip, Chinese cabbage, etc.), *Brassica napus* (rapeseed, etc.), *Raphanus sativus* (common radish), *Armoracia rusticana* (horseradish), *Matthiola* (stock) and the model organism *Arabidopsis thaliana* (thale cress).

Pieris rapae and other butterflies of the family Pieridae are some of the best-known pests of Brassicaceae species planted as commercial crops.

TAXONOMY

The family is included in the Brassicales according to the APG system. Older systems (e.g., Arthur Cronquist's) placed them into the Capparales, a now-defunct order that had a similar definition.

This family comprises about 375 genera and 2500 species all over the world; 94 species of 38 genera are found in Nepal. The plants are mostly herbs. A close relationship has long been acknowledged between the Brassicaceae and the caper family, Capparaceae, in part because members of both groups produce glucosinolate (mustard oil) compounds. The Capparaceae as traditionally circumscribed were paraphyletic with respect to Brassicaceae, with *Cleome* and several related genera being more closely related to the Brassicaceae than to other Capparaceae.The APG II system, therefore, has merged the two families under the name Brassicaceae. Other classifications have continued to recognize the Capparaceae, but with a more restricted circumscription, either including *Cleome* and its relatives in the Brassicaceae or recognizing them in the segregate family Cleomaceae. The APG III system has recently adopted this last solution, but this may change as a consensus arises on this point. This chapter deals with Brassicaceae *sensu stricto*, i.e. treating the Cleomaceae and Capparaceae as segregated families.

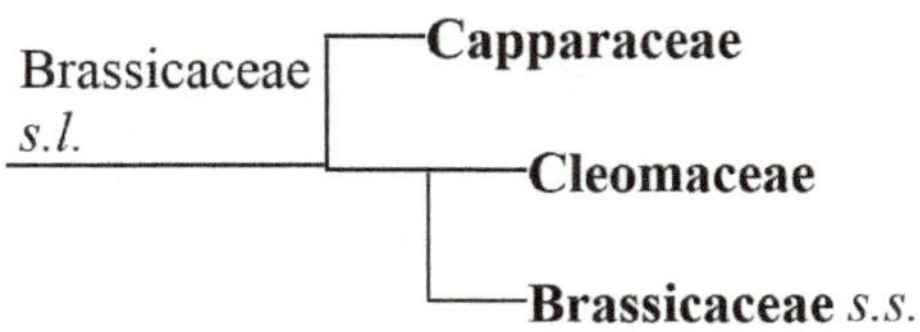

DESCRIPTION

The family consists mostly of herbaceous plants with annual, biennial, or perennial lifespans. However, around the Mediterranean, they include also a dozen woody shrubs 1-3 m tall, e.g. in northern Africa *(Zilla spinosa* and *Ptilotrichum spinosum)*, in the Dalmatian islands *(Dendralyssum* and *Cramboxylon)*, and chiefly in Canarias with some woody cruciferous genera: *Dendrosinapis, Descurainia, Parolinia, Stanleya, etc.*.

Aubrieta deltoidea (commonly known as purple rock cress) is a perennial wild flower used in gardening for its ornamental large inflorescence.

The leaves are alternate (rarely opposite), sometimes organized in basal rosettes; in rare shrubby crucifers of Mediterranean their leaves are mostly in terminal rosettes, and may be coriaceous and evergreen. They are very often pinnately incised and do not have stipules.

The structure of the flowers is extremely uniform throughout the family. They have four free saccate sepals and four clawed free petals, staggered. They can be disymmetric or slightly zygomorphic, with a typical cross-like arrangement (hence the name Cruciferae). They have six stamens, four of which are longer (as long as the petals) and are arranged in a cross like the petals and the other two are shorter (tetradynamous flower). The pistil is made up of two fused carpels and the style is very short, with two lobes. The ovary is superior. The flowers form ebracteate racemose inflorescences, often apically corymb-like.

Pollination occurs by entomogamy; nectar is produced at the base of the stamens and stored on the sepals.

Siliquae of *Cardamine impatiens*

The fruit is a peculiar kind of capsule named siliqua (plural siliquae). It opens by two valves, which are the modified carpels, leaving the seeds attached to a framework made up of the placenta and tissue from the junction between the valves (replum). Often, an indehiscent beak occurs at the top of the style and one or more seeds may be borne there. Where a siliqua is less than three times as long as it is broad, it is usually termed a silicula. The siliqua may break apart at constrictions occurring between the segments of the seeds, thus forming a sort of loment (e.g., *Raphanus*), it may eject the seeds explosively (e.g., *Cardamine*) or may be evolved in a sort of samara (e.g., *Isatis*). The fruit is often the most important diagnostic character for plants in this family. Most members share a suite of glucosinolate compounds that have a typical pungent odour usually associated with cole crops.

Uses

The importance of this family for food crops has led to its selective breeding throughout history. Some examples of cruciferous food plants are the cabbage, broccoli, cauliflower, turnip, rapeseed, mustard, radish, horseradish, cress, wasabi, and watercress.

Lunaria annua with ripe seed pods

Smelowskia americana is endemic to the midlatitude mountains of western North America.

Matthiola (stock), *Cheiranthus*, *Lobularia*, and *Iberis* (candytufts) are appreciated for their flowers. *Lunaria* (honesty) is cultivated for the decorative value of the translucent replum of the round silicula that remains on the dried stems after dehiscence.

Capsella bursa-pastoris, *Lepidium*, and many *Cardamine* species are common weeds.

Isatis tinctoria (woad) was used in the past to produce the colour indigo.

Arabidopsis thaliana is a very important model organism in the study of the flowering plants (Angiospermae).

CAULIFLOWER

Cauliflower is one of several vegetables in the species *Brassica oleracea* in the genus Brassica, which is in the family Brassicaceae. It is an annual plant that reproduces by seed. Typically, only the head (the *white curd*) is eaten. The cauliflower head is composed of a white inflorescence meristem. Cauliflower heads resemble those in broccoli, which differs in having flower buds. *Brassica oleracea* also includes broccoli, brussels sprouts, cabbage, collard greens, and kale, though they are of different cultivar groups.

History

The oldest record of cauliflower dates back to the 6th century B.C. In the 1st century AD, Pliny included what he called *cyma* among his descriptions of cultivated plants in *Natural History*: "*Ex omnibus brassicae generibus suavissima est cyma,*" ("Of all the varieties of cabbage the most pleasant-tasted is *cyma*"). Pliny's descriptions likely refer to the flowering heads of an earlier cultivated variety of *Brassica oleracea*, but comes close to describing modern cauliflower. In the 12th century, three varieties were described in Spain as introductions from Syria, where it had doubtless been grown for more than a thousand years. It is found in the writings of the Arab botanists Ibn al-'Awwam and Ibn al-Baitar, in the 12th and 13th centuries when its origins were said to be Cyprus.

François Pierre La Varenne employed *chouxfleurs* in *Le cuisinier françois*. They were introduced to France from Genoa in the 16th century, and are featured in Olivier de Serres' *Théâtre de l'agriculture* (1600), as *cauli-fiori* "as the Italians call it, which are still rather rare in France; they hold an honorable place in the garden because of their delicacy", but they did not commonly appear on grand tables until the time of Louis XIV. It was introduced in India in 1822 from England by the British.

Etymology

The origin of the name is from the Latin word *caulis* (cabbage) and flower.

Classification and Identification

Major Groups

There are four major groups of cauliflower.

Italian

This specimen is diverse in appearance, biennial and annual in type. This group also includes white, Romanesco, various brown, green, purple, and yellow cultivars. This type is the ancestral form from which the others were derived.

Northern European annuals

Used in Europe and North America for summer and fall harvest, it was developed in Germany in the 18th century, and includes the old cultivars Erfurt and Snowball.

Northwest European biennial

Used in Europe for winter and early spring harvest, this was developed in France in the 19th century, and includes the old cultivars Angers and Roscoff.

Asian

A tropical cauliflower used in China and India, it was developed in India during the 19th century from the now-abandoned Cornish type, and includes old varieties Early Benaras and Early Patna.

Varieties

There are hundreds of historic and current commercial varieties used around the world. A comprehensive list of about 80 North American varieties is maintained at North Carolina State University.

Colors

White

White cauliflower is the most common color of cauliflower.

Orange

Orange cauliflower (*B. oleracea* L. var. *botrytis*) contains 25% more vitamin A than white varieties. This trait originated from a natural mutant found in a cauliflower field in Canada. Cultivars include 'Cheddar' and 'Orange Bouquet'.

Green

Green cauliflower, of the *B. oleracea* botrytis group, is sometimes called broccoflower. It is available in the normal curd shape and with a fractal spiral curd called Romanesco broccoli. Both have been commercially available in the U.S. and Europe since the early 1990s. Green-curded varieties include 'Alverda', 'Green Goddess' and 'Vorda'. Romanesco varieties include 'Minaret' and 'Veronica'.

Purple

The purple color in this cauliflower is caused by the presence anthocyanins, water-soluble pigments that are found in many other plants and plant-based products, such as red cabbage and red wine. Varieties include 'Graffiti' and 'Purple Cape'.

In Great Britain and southern Italy, a broccoli with tiny flower buds is sold as a vegetable under the name "purple cauliflower"; it is not the same as standard cauliflower with a purple curd.

White cauliflower

Orange cauliflower

Purple cauliflower

Production

Major producers of cauliflowers (combined with broccoli) in millions of tonnes			
Country	Production		
People's Republic of China	9.1		
India	7.9		
Spain	0.5		

Mexico	0.5		
Italy	0.4		
World	22.3		

In 2013, global production of cauliflowers (combined for production reports with broccoli) was 22.3 million tonnes, with China and India together accounting for 76% of the total (table). Secondary producers, each having about 0.5 million tonnes annually, were Spain, Mexico and Italy (table).

Nutrition

Cauliflower, raw	
Nutritional value per 100 g (3.5 oz)	
Energy	104 kJ (25 kcal)
Carbohydrates	5 g
Sugars	1.9 g
Dietary fiber	2 g
Fat	0.3 g
Protein	1.9 g
Vitamins	
Thiamine (B1)	(4%) 0.05 mg
Riboflavin (B2)	(5%) 0.06 mg
Niacin (B3)	(3%) 0.507 mg
Pantothenic acid (B5)	(13%) 0.667 mg
Vitamin B6	(14%) 0.184 mg
Folate (B9)	(14%) 57 μg

Vitamin C	(58%) 48.2 mg
Vitamin E	(1%) 0.08 mg
Vitamin K	(15%) 15.5 μg
Minerals	
Calcium	(2%) 22 mg
Iron	(3%) 0.42 mg
Magnesium	(4%) 15 mg
Manganese	(7%) 0.155 mg
Phosphorus	(6%) 44 mg
Potassium	(6%) 299 mg
Sodium	(2%) 30 mg
Zinc	(3%) 0.27 mg
Other constituents	
Water	92 g
Link to USDA Database entry	
• Units • μg = micrograms • mg = milligrams • IU = International units	
Percentages are roughly approximated using US recommendations for adults. Source: USDA Nutrient Database	

100 grams of raw white cauliflower provides 25 calories, is low in fat, carbohydrates, dietary fiber and protein (table). It has a high content (20% or more of the Daily Value, DV) of vitamin C and moderate levels (10-19% DV) of several B vitamins and vitamin K (table).

Phytochemicals

Cauliflower contains several phytochemicals, common in the cabbage family, that are under preliminary research for their potential properties, including isothiocyanates and glucosinolates.

Boiling reduces the levels of cauliflower compounds, with losses of 20–30% after five minutes, 40–50% after ten minutes, and 75% after thirty minutes. However, other preparation methods, such as steaming, microwaving, and stir frying, have no significant effect on the compounds.

Cooking

Cauliflower can be roasted, boiled, fried, steamed, pickled, or eaten raw. When cooking, the outer leaves and thick stalks are removed, leaving only the florets. The leaves are also edible, but are most often discarded. The florets should be broken into similar-sized pieces so they are cooked evenly. After eight minutes of steaming, or five minutes of boiling, the florets should be soft, but not mushy (depending on size). Stirring while cooking can break the florets into smaller, uneven pieces.

Low carbohydrate dieters can use cauliflower as a reasonable substitute for potatoes or rice; while they can produce a similar texture, or mouth feel, they lack the starch of the originals. Like certain legumes (including chickpeas), it can be turned into a flour from which such foods as pizza or biscuits are made.

Aloo gobi, an Indian dish prepared with cauliflower and potato

Cooked cauliflower

Fractal Dimension

Cauliflower has been noticed by mathematicians for its distinct fractal dimension, predicted to be about 2.8. One of the fractal properties of cauliflower is that every branch, or "module", is similar to the entire cauliflower. Another quality, also present in other plant species, is that the angle between "modules," as they become more distant from the center, is 360 degrees divided by the golden ratio.

Fractal pattern of Romanesco broccoli, a variant form of cauliflower

CABBAGE

Cabbage or headed cabbage (comprising several cultivars of *Brassica oleracea*) is a leafy green or purple biennial plant, grown as an annual vegetable crop for its dense-leaved heads. It is descended from the wild cabbage, *B. oleracea* var. *oleracea*, and is closely related to broccoli and cauliflower (var. *botrytis*), brussels sprouts (var. *gemmifera*) and savoy cabbage (var. *sabauda*). Cabbage heads generally range from 0.5 to 4 kilograms (1 to 9 lb), and can be green, purple and white. Smooth-leafed firm-headed green cabbages are the most common, with smooth-leafed red and crinkle-leafed savoy cabbages of both colors seen more rarely. It is a multi-layered vegetable. Under conditions of long sunlit days such as are found at high northern latitudes in summer, cabbages can grow much larger. Some records are discussed at the end of the history section.

Cabbage and its cross section on a white background. Taken in Victoria, Australia in Jan 2009

It is difficult to trace the exact history of cabbage, but it was most likely domesticated somewhere in Europe before 1000 BC, although savoys were not developed until the 16th century. By the Middle Ages, it had become a prominent part of European cuisine. Cabbage heads are generally picked during the first year of the plant's life cycle, but plants intended for seed are allowed to grow a second year, and must be kept separated from other cole crops to prevent cross-pollination. Cabbage is prone to several nutrient deficiencies, as well as to multiple pests, and bacterial and fungal diseases.

The Food and Agriculture Organization of the United Nations (FAO) reports that world production of cabbage and other brassicas for 2011 was almost 69 million metric tons (68 million long tons; 75 million short tons). Almost half of these crops were grown in China, where Chinese cabbage is the most popular *Brassica* vegetable. Cabbages are prepared in many different ways for eating. They can be pickled, fermented for dishes such as sauerkraut, steamed, stewed, sautéed, braised, or eaten raw. Cabbage is a good source of vitamin K, vitamin C and dietary fiber. Contaminated cabbage has been linked to cases of food-borne illness in humans.

Description

Cabbage seedlings have a thin taproot and cordate (heart-shaped) cotyledon. The first leaves produced are ovate (egg-shaped) with a lobed petiole. Plants are 40–60 cm (16–24 in) tall in their first year at the mature vegetative stage, and 1.5–2.0 m (4.9–6.6 ft) tall when flowering in the second year. Heads average between 1 and 8 pounds (0.5 and 4 kg), with fast-growing, earlier-maturing varieties producing smaller heads. Most cabbages have thick, alternating leaves, with margins that range from wavy or lobed to highly dissected; some varieties have a waxy bloom on the leaves. Plants have root systems that are fibrous and shallow. About 90 percent of the root mass is in the upper 20–30 cm (8–12 in) of soil; some lateral roots can penetrate up to 2 m (6.6 ft) deep.

The cabbage inflorescence, which appears in the plant's second year of growth, features white or yellow flowers, each with four perpendicularly arranged petals.

The inflorescence is an unbranched and indeterminate terminal raceme measuring 50–100 cm (20–40 in) tall,with flowers that are yellow or white. Each flower has four petals set in a perpendicular pattern, as well as four sepals, six stamens, and a superior ovary that is two-celled and contains a single stigma and style. Two of the six stamens have shorter filaments. The fruit is a silique that opens at maturity through dehiscence to reveal brown or black seeds that are small and round in shape. Self-pollination is impossible, and plants are cross-pollinated by insects. The initial leaves form a rosette shape comprising 7 to 15 leaves, each measuring 25–35 cm (10–14 in) by 20–30 cm (8–12 in); after this, leaves with shorter petioles develop and heads form through the leaves cupping inward.

Many shapes, colors and leaf textures are found in various cultivated varieties of cabbage. Leaf types are generally divided between crinkled-leaf, loose-head savoys and smooth-leaf firm-head cabbages, while the color spectrum includes white and a range of greens and purples. Oblate, round and pointed shapes are found.

Cabbage has been selectively bred for head weight and morphological characteristics, frost hardiness, fast growth and storage ability. The appearance of the cabbage head has been given importance in selective breeding, with varieties being chosen for shape, color, firmness and other physical characteristics. Breeding objectives are now focused on increasing resistance to various insects and diseases and improving the nutritional content of cabbage. Scientific research into the genetic modification of *B. oleracea* crops, including cabbage, has included European Union and United States explorations of greater insect and herbicide resistance. Genetically modified *B. oleracea* crops are not currently used in commercial agriculture.

Cultivation

Cabbage is generally grown for its densely leaved heads, produced during the first year of its biennial cycle. Plants perform best when grown in well-drained soil in a location that receives full sun. Different varieties prefer different soil types, ranging from lighter sand to heavier clay, but all prefer fertile ground with a pH between 6.0 and 6.8. For optimal growth, there must be adequate levels of nitrogen in the soil, especially during the early head formation stage, and sufficient phosphorus and potassium during the early stages of expansion of the outer leaves. Temperatures between 4 and 24 °C (39 and 75 °F) prompt the best growth, and extended periods of higher or lower temperatures may result in premature bolting (flowering).

A cabbage field

Flowering induced by periods of low temperatures (a process called vernalization) only occurs if the plant is past the juvenile period. The transition from a juvenile to adult state happens when the stem diameter is about 6 mm (0.24 in). Vernalization allows the plant to grow to an adequate size before flowering. In certain climates, cabbage can be planted at the beginning of the cold period and survive until a later warm period without being induced to flower, a practice that was common in the eastern US.

Green and purple cabbages

Plants are generally started in protected locations early in the growing season before being transplanted outside, although some are seeded directly into the ground from which they will be harvested. Seedlings typically emerge in about 4–6 days from seeds planted 1.3 cm (0.5 in) deep at a soil temperature between 20 and 30 °C (68 and 86 °F). Growers normally place plants 30 to 61 cm (12 to 24 in) apart. Closer spacing reduces the resources available to each plant (especially the amount of light) and increases the time taken to reach maturity. Some varieties of cabbage have been developed for ornamental use; these are generally called «flowering cabbage». They do not produce heads and feature purple or green outer leaves surrounding an inner grouping of smaller leaves in white, red, or pink. Early varieties of cabbage take about 70 days from planting to reach maturity, while late varieties take about 120 days. Cabbages are mature when they are firm and solid to the touch. They are harvested by cutting the stalk just below the bottom leaves with a blade. The outer leaves are trimmed, and any diseased, damaged, or necrotic leaves are removed. Delays in harvest can result in the head splitting as a result of expansion of the inner leaves and continued stem growth. Factors that contribute to reduced head weight include: growth in the compacted soils that result from no-till farming practices, drought, waterlogging, insect and disease incidence, and shading and nutrient stress caused by weeds.

When being grown for seed, cabbages must be isolated from other *B. oleracea* subspecies, including the wild varieties, by 0.8 to 1.6 km (0.5 to 1 mi) to prevent cross-pollination. Other *Brassica* species, such as *B. rapa*, *B. juncea*, *B. nigra*, *B. napus* and *Raphanus sativus*, do not readily cross-pollinate.

Cultivars

There are several cultivar groups of cabbage, each including many cultivars:

- Savoy – Characterized by crimped or curly leaves, mild flavor and tender texture
- Spring Greens – Loose-headed, commonly sliced and steamed
- Green – Light to dark green, slightly pointed heads. This is the most commonly grown cultivar.
- Red – Smooth red leaves, often used for pickling or stewing
- White, also called Dutch – Smooth, pale green leaves

White cabbage

Some sources only delineate three cultivars: savoy, red and white, with spring greens and green cabbage being subsumed into the latter.

Cultivation Problems

Due to its high level of nutrient requirements, cabbage is prone to nutrient deficiencies, including boron, calcium, phosphorus and potassium.There are several physiological disorders that can affect the postharvest appearance of cabbage. Internal tip burn occurs when the margins of inside leaves turn brown, but the outer leaves look normal. Necrotic spot is where there are oval sunken spots a few millimeters across that are often grouped around the midrib. In pepper spot, tiny black spots occur on the areas between the veins, which can increase during storage.

Fungal diseases include wirestem, which causes weak or dying transplants; *Fusarium* yellows, which result in stunted and twisted plants with yellow leaves; and blackleg (see *Leptosphaeria maculans*), which leads to sunken areas on stems and gray-brown spotted leaves. The fungi *Alternaria brassicae* and *A. brassicicola* cause dark leaf spots in affected plants. They are both seedborne and airborne, and typically propagate from spores in infected plant debris left on the soil surface for up to twelve weeks after harvest. *Rhizoctonia solani* causes the post-emergence disease wirestem, resulting in killed seedlings ("damping-off"), root rot or stunted growth and smaller heads.

Cabbage moth damage to a savoy cabbage

One of the most common bacterial diseases to affect cabbage is black rot, caused by *Xanthomonas campestris*, which causes chlorotic and necrotic lesions that start at the leaf margins, and wilting of plants. Clubroot, caused by the soilborne slime mold-like organism *Plasmodiophora brassicae*, results in swollen, club-like roots. Downy mildew, a parasitic disease caused by the oomycete *Peronospora parasitica*,produces pale leaves with white, brownish or olive mildew on the lower leaf surfaces; this is often confused with the fungal disease powdery mildew.

Pests include root-knot nematodes and cabbage maggots, which produce stunted and wilted plants with yellow leaves; aphids, which induce stunted plants with curled and yellow leaves; harlequin bugs, which cause white and yellow leaves; thrips, which lead to leaves with white-bronze spots; striped flea beetles, which riddle leaves with small holes; and caterpillars, which leave behind large, ragged holes in leaves. The caterpillar stage of the "small cabbage white butterfly" (*Pieris rapae*), commonly known in the United

States as the "imported cabbage worm", is a major cabbage pest in most countries. The large white butterfly (*Pieris brassicae*) is prevalent in eastern European countries. The diamondback moth (*Plutella xylostella*) and the cabbage moth (*Mamestra brassicae*) thrive in the higher summer temperatures of continental Europe, where they cause considerable damage to cabbage crops. In India, the diamondback moth has caused losses up to 90 percent in crops that were not treated with insecticide. Destructive soil insects include the cabbage root fly (*Delia radicum*) and the cabbage maggot (*Hylemya brassicae*), whose larvae can burrow into the part of plant consumed by humans.

Planting near other members of the cabbage family, or where these plants have been placed in previous years, can prompt the spread of pests and disease. Excessive water and excessive heat can also cause cultivation problems.

Production

The Food and Agriculture Organization of the United Nations (FAO) combines cabbage and other brassicas for reporting purposes. Total world production of all brassicas for calendar year 2012 was 70,104,972 metric tons (68,997,771 long tons; 77,277,504 short tons). The nations with the largest production were China, which produced 47 percent of the world total, and India, which produced 12 percent. China and India used a surface area of 980,000 hectares (2,400,000 acres) and 375,000 hectares (930,000 acres), respectively, to grow these crops; the total global surface area used for cabbage and related *Brassica* crops in 2012 was 2,391,747 hectares (5,910,140 acres). The largest yields were from South Korea, which harvested 71,188.6 kilograms per hectare, Ireland (68,888.9 kg/ha), and Japan (67,647.1 kg/ha).

Top ten cabbage and other brassica producersin 2012

Country	**Production** (tonnes)	**Source**
China	32,800,000	FAO estimate
India	8,500,000	FAO estimate
Russia	3,309,315	official figure
Japan	2,300,000	FAO estimate
South Korea	2,118,930	official figure
Ukraine	1,922,400	official figure
Indonesia	1,487,531	official figure
Poland	1,198,726	official figure
Romania	990,154	official figure
United States	964,830	official figure
World	**70,104,972**	aggregate

Cabbages sold for market are generally smaller, and different varieties are used for those sold immediately upon harvest and those stored before sale. Those used for processing, especially sauerkraut, are larger and have a lower percentage of water. Both hand and mechanical harvesting are used, with hand-harvesting generally used for cabbages destined for market sales. In commercial-scale operations, hand-harvested cabbages are trimmed, sorted, and packed directly in the field to increase efficiency. Vacuum cooling rapidly refrigerates the vegetable, allowing for earlier shipping and a fresher product. Cabbage can be stored the longest at −1 to 2 °C (30 to 36 °F) with a humidity of 90–100 percent; these conditions will result in up to six months of longevity. When stored under less ideal conditions, cabbage can still last up to four months.

BRUSSELS SPROUT

The Brussels sprout is a member of the Gemmifera Group of cabbages (*Brassica oleracea*), grown for its edible buds.

The leafy green vegetables are typically 2.5–4 cm (0.98–1.6 in) in diameter and look like miniature cabbages. The Brussels sprout has long been popular in Brussels, Belgium, and may have originated and gained its name there.

Cultivation

Forerunners to modern Brussels sprouts were likely cultivated in Ancient Rome. Brussels sprouts as they are now known were grown possibly as early as the 13th century in what is now Belgium. The first written reference dates to 1587. During the 16th century, they enjoyed a popularity in the Southern Netherlands that eventually spread throughout the cooler parts of Northern Europe.

Brussels sprouts grow in temperature ranges of 7–24 °C (45–75 °F), with highest yields at 15–18 °C (59–64 °F). Fields are ready for harvest 90 to 180 days after planting. The edible sprouts grow like buds in helical patterns along the side of long, thick stalks of about 60 to 120 cm (24 to 47 in) in height, maturing over several weeks from the lower to the upper part of the stalk. Sprouts may be picked by hand into baskets, in which case several harvests are made of five to 15 sprouts at a time, or by cutting the entire stalk at once for processing, or by mechanical harvester, depending on variety. Each stalk can produce 1.1 to 1.4 kg (2.4 to 3.1 lb), although the commercial yield is about 900 g (2.0 lb) per stalk. Harvest season in temperate zones of the northern latitudes is September to March, making Brussels sprout a traditional winter stock vegetable. In the home garden, harvest can be delayed as quality does not suffer from freezing. Sprouts are considered to be sweetest after a frost.

Brussels sprouts are a cultivar group of the same species as cabbage, in the same family as collard greens, broccoli, kale, and kohlrabi; they are cruciferous (they belong to the Brassicaceae family; old name Cruciferae). Many cultivars are available, some being purple in colour, such as ‹Ruby Crunch› or ‹Red Bull›.

Europe

In Continental Europe, the largest producers are the Netherlands, at 82,000 metric tons, and Germany, at 10,000 tons. The United Kingdom has production comparable to that of the Netherlands, but it is not generally exported.

North America

Production of Brussels sprouts in the United States began in the 18th century, when French settlers brought them to Louisiana. The first plantings in California's Central Coast began in the 1920s, with significant production beginning in the 1940s. Currently, several thousand acres are planted in coastal areas of San Mateo, Santa Cruz, and Monterey counties of California, which offer an ideal combination of coastal fog and cool temperatures year-round. The harvest season lasts from June through January.

Most American production is in California, with a smaller percentage of the crop grown in Skagit Valley, Washington, where cool springs, mild summers, and rich soil abounds, and to a lesser degree on Long Island, New York. Total American production is approximately 32,000 tons, with a value of $27 million.

About 80% to 85% of US production is for the frozen food market, with the remainder for fresh consumption. Once harvested, sprouts last three to five weeks under ideal near-freezing conditions before wilting and discolouring, and about half as long at refrigerator temperature. American varieties are generally 2.5–5 cm (0.98–1.97 in) in diameter.

Ontario, Canada, produces about 1,000 tons per year.

They are also grown in Baja California, Mexico, where the harvest season is from December through June.

Nutrients, Phytochemicals and Research

Raw Brussels sprouts contain excellent levels of vitamin C and vitamin K, with more moderate amounts of B vitamins, such as folic acid and vitamin B6 (USDA nutrient table, right); essential minerals and dietary fibre exist in lesser amounts (table).

Brussels sprouts, as with broccoli and other brassicas, contain sulforaphane, a phytochemical under basic research for its potential anticancer properties. Although boiling reduces the level of sulforaphane, steaming and stir frying do not result in significant loss.

Consuming Brussels sprouts in excess may not be suitable for patients taking anticoagulants such as warfarin since they contain vitamin K, a blood-clotting factor. In one such reported incident, eating too many Brussels sprouts may have countered blood-thinning therapy.

Cooking and Preparation

The most common method of preparing Brussels sprouts for cooking begins with cutting the buds off the stalk. Any surplus stem is cut away, and any loose surface leaves are peeled and discarded. Once cut and cleaned, the buds are typically cooked by boiling, steaming, stir frying, grilling, or roasting. To ensure even cooking throughout, buds of a similar size are usually chosen. Some cooks will make a single cut or a cross in the center of the stem to aid the penetration of heat. Brussels sprouts can be pickled as an alternative to cooking.

Overcooking will render the buds gray and soft, and they then develop a strong flavor and odor that some dislike. The odor is associated with the glucosinolate sinigrin, an organic compound that contains sulfur: hence the strong smell. For taste, roasting Brussels sprouts is a common way to cook them to bring out flavor. Common toppings or additions for Brussels sprouts include Parmesan cheese and butter, balsamic vinegar, apple cider vinegar, bacon, pistachios, pine nuts, mustard, brown sugar, and pepper. Another popular way of cooking Brussels sprouts is to sauté them.

BROCCOLI

Broccoli is an edible green plant in the cabbage family whose large flowering head is eaten as a vegetable.

The word *broccoli* comes from the Italian plural of *broccolo*, which means "the flowering crest of a cabbage", and is the diminutive form of *brocco*, meaning "small nail" or "sprout". Broccoli is often boiled or steamed but may be eaten raw.

Broccoli is classified in the Italica cultivar group of the species *Brassica oleracea*. Broccoli has large flower heads, usually green in color, arranged in a tree-like structure branching out from a thick, edible stalk. The mass of flower heads is surrounded by leaves. Broccoli resembles cauliflower, which is a different cultivar group of the same species.

Broccoli and its cross section isolated on a white background.

Broccoli is a result of careful breeding of cultivated *Brassica* crops in the northern Mediterranean starting in about the 6th century BC. Since the time of the Roman Empire, broccoli has been considered a uniquely valuable food among Italians. Broccoli was brought to England from Antwerp in the mid-18th century by Peter Scheemakers. Broccoli was first introduced to the United States by Southern Italian immigrants, but did not become widely popular until the 1920s.

Varieties

There are three commonly grown types of broccoli. The most familiar is Calabrese broccoli, often referred to simply as "broccoli", named after Calabria in Italy. It has large (10 to 20 cm) green heads and thick stalks. It is a cool season annual crop. Sprouting broccoli has a larger number of heads with many thin stalks. Purple cauliflower is a type of broccoli sold in southern Italy, Spain, and the United Kingdom. It has a head shaped like cauliflower, but consisting of tiny flower buds. It sometimes, but not always, has a purple cast to the tips of the flower buds.

Broccoli plants in a nursery

Other cultivar groups of *Brassica oleracea* include cabbage (Capitata Group), cauliflower and Romanesco broccoli (Botrytis Group), kale and collard greens (Acephala Group), kohlrabi (Gongylodes Group), Brussels sprouts (Gemmifera Group), and kai-lan (Alboglabra Group). Rapini, sometimes called "broccoli raab" among other names, forms similar but smaller heads, and is actually a type of turnip (*Brassica rapa*). Broccolini or "Tenderstem broccoli" is a cross between broccoli and Chinese broccoli. Beneforté is a variety of broccoli containing 2–3 times more glucoraphanin that was produced by crossing broccoli with a wild Brassica variety, *Brassica oleracea* var *villosa*.

Production

Major producers of broccoli (combined with cauliflowers) in millions of tonnes			
Country	**Production**		
People's Republic of China	9.1		
India	7.9		
Spain	0.5		
Mexico	0.5		
Italy	0.4		
World	22.3		

In 2013, global production of broccoli (combined for production reports with cauliflowers) was 22.3 million tonnes, with China and India together accounting for 76% of the total (table). Secondary producers, each having about 0.5 million tonnes annually, were Spain, Mexico and Italy (table).

Cultivation

Broccoli is a cool-weather crop that does poorly in hot summer weather. Broccoli grows best when exposed to an average daily temperature between 18 and 23 °C (64 and 73 °F). When the cluster of flowers, also referred to as a «head» of broccoli, appear in the center of the plant, the cluster is green. Garden pruners or shears are used to cut the head about an inch from the tip. Broccoli should be harvested before the flowers on the head bloom bright yellow.

While the heading broccoli variety performs poorly in hot weather, mainly due to insect infestation, the sprouting variety is more resistant, though attention must be paid to sucking insects (such as aphids), caterpillars and whiteflies. Spraying of *bacillus thuringiensis* can control caterpillar attacks, while a citronella vase may ward off whiteflies.

Pests

Mostly introduced by accident, "cabbage worms", the larvae of *Pieris rapae*, the small white butterfly are a common pest in broccoli.

REFERENCES

1. Angiosperm Phylogeny Group (2009). "An update of the Angiosperm Phylogeny Group classification for the orders and families of flowering plants: APG III" (PDF). Botanical Journal of the Linnean Society. 161 (2): 105–121. doi:10.1111/j.1095-8339.2009.00996.x. Retrieved 2013-07-06.
2. "Brassicaceae". The Plant List.
3. Hall, J. C.; Sytsma, K. J.; Iltis, H. H. (2002). "Phylogeny of Capparaceae and Brassicaceae based on chloroplast sequence data". American Journal of Botany. 89 (11): 1826–1842. doi:10.3732/ajb.89.11.1826. PMID 21665611.
4. Merriam-Webster, Merriam-Webster's Unabridged Dictionary, Merriam-Webster.
5. Pliny (the Elder) (1841). Weise, C.H., ed. Historiae Naturalis Libri XX (in Latin). p. 249.
6. Rackham, H., ed. (1949). "XXXV". Pliny's Natural History. Retrieved 1 July 2015.
7. Crozier, Arthur Alger (1891). The Cauliflower. Ann Arbor, Michigan: Register Publishing Co. p. 12.
8. "Cabbage Flowers for Food". Aggie Horticulture. Texas AgriLife Extension Service, Texas A&M System. Retrieved 12 October 2014.
9. Fenwick, G. Roger; Heaney, Robert K.; Mullin, W. John; VanEtten, Cecil H. (1982). "Glucosinolates and their breakdown products in food and food plants". C R C Critical Reviews in Food Science and Nutrition. 18 (2): 123–201. doi:10.1080/10408398209527361.
10. "History of Cauliflower". cauliflowerfestival.com. Retrieved 12 October 2014.
11. Wheaton, Barbara Ketcham (1996) Savoring the Past: the French kitchen and table from 1300 to 1789, Touchstone, p. 118, ISBN 0-684-81857-4.
12. Wheaton, Barbara Ketcham (1996) Savoring the Past: the French kitchen and table from 1300 to 1789, Touchstone, p. 66, ISBN 0-684-81857-4.
13. Toussaint-Samat, Maguelonne (2009) A History of Food, 2nd ed., John Wiley & Sons, pp. 625f, ISBN 1-4443-0514-X.
14. Vegetable Crops. New India Publishing. 1 January 2007. p. 209. ISBN 978-81-89422-41-7.
15. "Cauliflower: definition". Reference.com. 2006. Retrieved 22 November 2008.
16. Crisp, P. (1982). "The use of an evolutionary scheme for cauliflowers in screening of genetic resources". Euphytica. 31 (3): 725. doi:10.1007/BF00039211.
17. Swarup, V.; Chatterjee, S.S (1972). "Origin and genetic improvement of Indian cauliflower". Economic Botany. 26 (4): 381–393. doi:10.1007/BF02860710.
18. Farnham, M. (2007). "Vegetable Cultivar Descriptions for North America:Cauliflower". Retrieved 2007-09-19.
19. Cohen, Amanda; Koenig, Leah (November 9, 2009). "The Story Behind Orange Cauliflower". saveur.com.

20. Dickson, M.H.; Lee C.Y.; Blamble A.E. (1988). "Orange-curd high carotene cauliflower inbreds, NY 156, NY 163, and NY 165". HortScience. 23: 778–779.
21. Chiu, L.; Prior, R.L.; Wu, X.; Li, L. (July 16, 2005). "Toward Identification of the Candidate Gene Controlling Anthocyanin Accumulation in Purple Cauliflower (Brassica oleracea L. var. botrytis)". American Society of Plant Biologists Annual Meeting. p. 628.
22. "Production/Crops, Quantities by Country for Cauliflowers and Broccoli for 2013". Food and Agricultural Organization of the United Nations, Statistics Division (FAOSTAT). Retrieved 23 November 2015.
23. "Cauliflower Nutrient Data, Raw per 100 g". USDA National Nutrient Database, Release 28. 2014. Retrieved 15 May 2013.
24. Ishida M, Hara M, Fukino N, Kakizaki T, Morimitsu Y (2014). "Glucosinolate metabolism, functionality and breeding for the improvement of Brassicaceae vegetables". Breed Sci. 64 (1): 48–59. doi:10.1270/jsbbs.64.48. PMC 4031110Freely accessible. PMID 24987290.
25. Warwick Medical School, University of Warwick (2007-05-15). "Research Says Boiling Broccoli Ruins Its Anti Cancer Properties.".
26. Stephens, MJ (1998). "Secondary Edible Parts of Vegetables". Vegetarian. 5.
27. Carlson, Jen (10 March 2014). "Making Pizza Dough Out Of Cauliflower Is A Delicious Revelation". Gothamist. Retrieved 5 February 2015.
28. "Top 10 Paleo Cauliflower Recipes". Empowered Sustenance. 9 September 2013. Retrieved 5 February 2015.
29. Walker, John. (2005-03-22) Fractal Food. Fourmilab.ch. Retrieved on 2013-09-03.
30. Description of the Julia sets of the cabbage fractal. Iwriteiam.nl. Retrieved on 2013-09-03.
31. Kim, Sang-Hoon. "Fractal Structure of a White Cauliflower" (PDF). Journal of Korean physical society. 46 (2): 474–477. arXiv:cond-mat/0409763Freely accessible.
32. Kim, Sang-Hoon (2004). "Fractal dimensions of a green broccoli and a white cauliflower". arXiv:cond-mat/0411597Freely accessible.
33. "Romanesco cauliflower is a striking example of fractals".

CHAPTER 4

Allium

Chapter Contents

- **Introduction**
- **Description**
- **Taxonomy**
- **Onion**
- **Garlic**
- **Shallot**
- **Chives**

Allium is a genus of perennial bulbous plants that produce chemical compounds (mostly cysteine sulfoxide) that give them a characteristic onion or garlic taste and odor. Many are used as food plants, though not all members of the genus are equally flavorful. In most cases, both bulb and leaves are edible. Their taste may be strong or weaker, depending on the species.' There are actually over 700 species with 57 genera. Most commonly known species include garlic, onion, leeks, chives, shallots, and ramps.

INTRODUCTION

Allium is a genus of monocotyledonous flowering plants that includes the cultivated onion, garlic, scallion, shallot and leek as well as chives and hundreds of other wild species.

The generic name *Allium* is the Latin word for garlic, and Linnaeus first described the genus *Allium* in 1753. Some sources refer to Greek aleo, to avoid by reason of the smell of garlic. The cooking and consumption of parts of the plants is due to the large variety of flavors and textures of the species. Various *Allium* have been cultivated from the earliest times and about a dozen species are economically important as crops, or garden vegetables, and an increasing number of species are important as ornamental plants. The inclusion of a species to the genus *Allium* is taxonomically difficult and species boundaries are unclear.

Estimates of the number of species have been as low as 260, and as high as 979, but are most likely about 800–850. The type species for the genus is *Allium sativum*.

Allium species occur in temperate climates of the Northern Hemisphere, except for a few species occurring in Chile (such as *A. juncifolium*), Brazil (*A. sellovianum*), and tropical Africa (*A. spathaceum*). They vary in height between 5 cm and 150 cm. The flowers form an umbel at the top of a leafless stalk. The bulbs vary in size between species, from small (around 2–3 mm in diameter) to rather large (8–10 cm). Some species (such as Welsh onion *A. fistulosum*) develop thickened leaf-bases rather than forming bulbs as such.

Plants of the *Allium* genus produce chemical compounds (mostly derived from cysteine sulfoxides) that give them a characteristic (alliaceous) onion or garlic taste and odor. Many are used as food plants, though not all members of the genus are equally flavorous. In most cases, both bulb and leaves are edible and the taste may be strong or weak, depending on the species and on ground sulfur (usually as sulfate) content. In the rare occurrence of sulfur-free growth conditions, all *Allium* species lose their usual pungency altogether.

In the APG III classification system, *Allium* is placed in the family Amaryllidaceae, subfamily Allioideae (formerly the family Alliaceae). In some of the older classification systems, *Allium* was placed in Liliaceae. Molecular phylogenetic studies have shown this circumscription of Liliaceae is not monophyletic.

Allium is one of about fifty-seven genera of flowering plants with more than 500 species. It is by far the largest genus in the Amaryllidaceae, and also in the Alliaceae in classification systems in which that family is recognized as separate.

DESCRIPTION

Allium species are herbaceous perennials with flowers produced on scapes. They grow from solitary or clustered tunicate bulbs and many have an onion odor and taste. Plants are perennialized by bulbs that reform annually from the base of the old bulb, or are produced on the ends of rhizomes or, in a few species, at the ends of stolons. A small number of species have tuberous roots. The bulbs' outer coats are commonly brown or grey, with a smooth texture, and are fibrous, or with cellular reticulation. The inner coats of the bulbs are membranous.

Capsule of *Allium oreophilum*.

Many alliums have basal leaves that commonly wither away from the tips downward before or while the plants flower, but some species have persistent foliage. Plants produce from one to 12 leaves, most species having linear, channeled or flat leaf blades. The leaf blades are straight or variously coiled, but some species have broad leaves, including *A. victorialis* and *A. tricoccum*. The leaves are sessile, and very rarely narrowed into a petiole.

The flowers are erect or in some species pendent, having six petal-like tepals produced in two whorls. The flowers have one style and six epipetalous stamens; the anthers and pollen can vary in color depending on the species. The ovaries are superior, and three-lobed with three locules.

The fruits are capsules that open longitudinally along the capsule wall between the partitions of the locule. The seeds are black, and have a rounded shape.

The terete or flattened flowering scapes are normally persistent. The inflorescences are umbels, in which the outside flowers bloom first and flowering progresses to the inside. Some species produce bulbils within the umbels, and in some species, such as *Allium paradoxum*, the bulbils replace some or all the flowers. The umbels are subtended by noticeable spathe bracts, which are commonly fused and normally have around three veins.

Some bulbous alliums increase by forming little bulbs or "offsets" around the old one, as well as by seed. Several species can form many bulbils in the flowerhead; in the so-called "tree onion" or Egyptian onion (*A.* ×*proliferum*) the bulbils are few, but large enough to be pickled.

Many of the species of *Allium* have been used as food items throughout their ranges. There are several poisonous species that are somewhat similar in appearance (e.g. in North America, death camas, *Toxicoscordion venenosum*), but none of these has the distinctive scent of onions or garlic.

TAXONOMY

With over 850 species Allium is the sole genus in the Allieae, one of four tribes of subfamily Allioideae (Amaryllidaceae). New species continue to be described and *Allium* is one of the largest monocotyledonous genera, but the precise taxonomy of *Allium* is poorly understood, with incorrect descriptions being widespread. The difficulties arise from the fact that the genus displays considerable polymorphism and has adapted to a wide variety of habitats. Furthermore, traditional classications had been based on homoplasious characteristics (the independent evolution of similar features in species of different lineages). However, the genus has been shown to be monophyletic, containing three major clades, although some proposed subgenera are not. Some progress is being made using molecular phylogenetic methods, and the internal transcribed spacer (ITS) region, including the 5.8S rDNA and the two spacers ITS1 and ITS2, is one of the more commonly used markers in the study of the differentiation of the *Allium* species.

Allium includes a number of taxonomic groupings previously considered separate genera (*Caloscordum* Herb., *Milula* Prain and *Nectaroscordum* Lindl.) *Allium spicatum* had been treated by many authors as *Milula spicata*, the only species in the monospecific genus *Milula*. In 2000, it was shown to be embedded in *Allium*.

Phylogeny

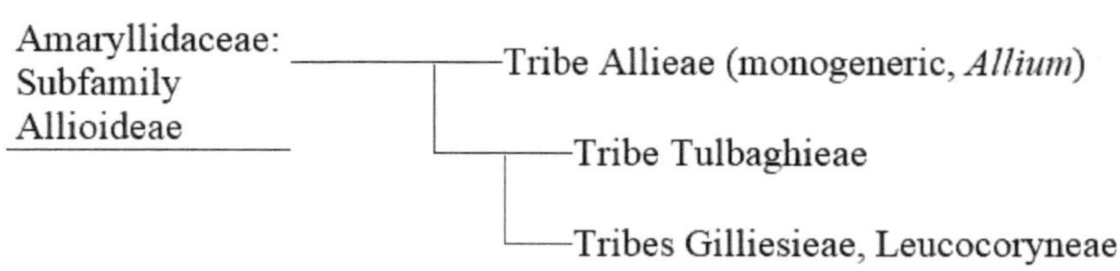

Distribution and Habitat

The majority of *Allium* species are native to the Northern Hemisphere, being spread throughout the holarctic region, from dry subtropics to the boreal zone, but predominantly in Asia. Of the latter 138 species occur in China, about a sixth of all species, representing five subgenera. A few species are native to Africa and Central and South America. A single known exception, *Allium dregeanum* occurs in the Southern Hemisphere (South Africa). There are two centres of diversity, a major one from the Mediterranean Basin to Central Asia and Pakistan, while a minor one is found in western North America.

Ecology

Species grow in various conditions from dry, well-drained mineral-based soils to moist, organic soils; most grow in sunny locations, but a number also grow in forests (e.g., *A. ursinum*), or even in swamps or water.

Various *Allium* species are used as food plants by the larvae of the leek moth and onion fly as well as some Lepidoptera including cabbage moth, common swift moth (recorded on garlic), garden dart moth, large yellow underwing moth, nutmeg moth, setaceous Hebrew character moth, turnip moth and *Schinia rosea*, a moth that feeds exclusively on *Allium* species.

Cultivation

Many *Allium* species have been harvested through human history, but only about a dozen are still economically important today as crops or garden vegetables.

Selection of cultivated alliums displayed at the *BBC Gardeners' World* show

Ornamental

Many *Allium* species and hybrids are cultivated as ornamentals. These include *A. cristophii* and *A. giganteum*, which are used as border plants for their ornamental flowers, and their "architectural" qualities. Several hybrids have been bred, or selected, with rich purple flowers. *A. hollandicum* 'Purple Sensation' is one of the most popular and has been given an Award of Garden Merit (H4). These ornamental onions produce spherical umbels on single stalks in spring and summer, in a wide variety of sizes and colours, ranging from white (*Allium* 'Mont Blanc'), blue (*A. caeruleum*), to yellow (*A. flavum*) and purple (*A. giganteum*). By contrast, other species (such as invasive *A. triquetrum* and *A. ursinum*) can become troublesome garden weeds. The hybrid cultivars 'Beau Regard', 'Gladiator', and 'Globemaster' have gained the Royal Horticultural Society's Award of Garden Merit.

Toxicity

Dogs and cats are very susceptible to poisoning after the consumption of certain species.

Uses

The genus includes many economically important species. These include onions (*A. cepa*), French shallots (*A. oschaninii*), leeks (*A. ampeloprasum*), scallions (various *Allium* species), and herbs such as garlic (*A. sativum*) and chives (*A. schoenoprasum*). Some have been used as traditional medicines.

ONION

The onion (*Allium cepa* L., from Latin *cepa* "onion"), also known as the bulb onion or common onion, is a vegetable and is the most widely cultivated species of the genus *Allium*.

This genus also contains several other species variously referred to as onions and cultivated for food, such as the Japanese bunching onion (*Allium fistulosum*), the tree onion (*A. ×proliferum*), and the Canada onion (*Allium canadense*). The name "wild onion" is applied to a number of *Allium* species, but *A. cepa* is exclusively known from cultivation. Its ancestral wild original form is not known, although escapes from cultivation have become established in some regions. The onion is most frequently a biennial or a perennial plant, but is usually treated as an annual and harvested in its first growing season.

The onion plant has a fan of hollow, bluish-green leaves and its bulb at the base of the plant begins to swell when a certain day-length is reached. In the autumn (or in spring, in the case of overwintering onions), the foliage dies down and the outer layers of the bulb become dry and brittle. The crop is harvested and dried and the onions are ready for use or storage. The crop is prone to attack by a number of pests and diseases, particularly the onion fly, the onion eelworm, and various fungi cause rotting. Some varieties of *A. cepa*, such as shallots and potato onions, produce multiple bulbs.

Mixed onions: red and brown onions, with and without skin, whole and sliced and in rings.

Onions are cultivated and used around the world. As a food item, they are usually served cooked, as a vegetable or part of a prepared savoury dish, but can also be eaten raw or used to make pickles or chutneys. They are pungent when chopped and contain certain chemical substances which irritate the eyes.

Taxonomy and Etymology

The onion plant (*Allium cepa*), also known as the bulb onion or common onion, is the most widely cultivated species of the genus *Allium.* It was first officially described by Carl Linnaeus in his 1753 work *Species Plantarum.* A number of synonyms have appeared in its taxonomic history:

- *Allium cepa* var. *aggregatum* – G. Don
- *Allium cepa* var. *bulbiferum* – Regel
- *Allium cepa* var. *cepa* – Linnaeus
- *Allium cepa* var. *multiplicans* – L.H. Bailey
- *Allium cepa* var. *proliferum* – (Moench) Regel
- *Allium cepa* var. *solaninum* – Alef
- *Allium cepa* var. *viviparum* – (Metz) Mansf.

A. cepa is known exclusively from cultivation, but related wild species occur in Central Asia. The most closely related species include *A. vavilovii* (Popov & Vved.) and *A. asarense* (R.M. Fritsch & Matin) from Iran. However, Zohary and Hopf state that "there are doubts whether the *A. vavilovii* collections tested represent genuine wild material or only feral derivatives of the crop."

The vast majority of cultivars of *A. cepa* belong to the "common onion group" (*A. cepa* var. *cepa*) and are usually referred to simply as "onions". The Aggregatum group of cultivars (*A. cepa* var. *aggregatum*) includes both shallots and potato onions.

The genus *Allium* also contains a number of other species variously referred to as onions and cultivated for food, such as the Japanese bunching onion (*A. fistulosum*), Egyptian onion (*A. ×proliferum*), and Canada onion (*A. canadense*).

Cepa is commonly accepted as Latin for "onion" and has an affinity with Ancient Greek: κάπια (*kápia*), Albanian: *qepë*, Aromanian: *tseapã*, Catalan: *ceba*, English: chive, Occitan: *ceba*, Old French: *cive*, and Romanian: *ceap* .

Description

The onion plant has been grown and selectively bred in cultivation for at least 7,000 years. It is a biennial plant, but is usually grown as an annual. Modern varieties typically grow to a height of 15 to 45 cm (6 to 18 in). The leaves are yellowish- to bluish green and grow alternately in a flattened, fan-shaped swathe. They are fleshy, hollow, and cylindrical, with one flattened side. They are at their broadest about a quarter of the way up, beyond which they taper towards a blunt tip. The base of each leaf is a flattened, usually white sheath that grows out of a basal disc. From the underside of the disc, a bundle of fibrous roots extends for a short way into the soil. As the onion matures, food reserves begin to accumulate in the leaf bases and the bulb of the onion swells.

Stereo image

Left frame

Right frame

Parallel view (⧠)

Cross-eye view (☒)

Onion seeds have a very distinct shape.

In the autumn, the leaves die back and the outer scales of the bulb become dry and brittle, so the crop is then normally harvested. If left in the soil over winter, the growing point in the middle of the bulb begins to develop in the spring. New leaves appear and a long, stout, hollow stem expands, topped by a bract protecting a developing inflorescence. The inflorescence takes the form of a globular umbel of white flowers with parts in sixes. The seeds are glossy black and triangular in cross section.

Cultivation

Onions are best cultivated in fertile soils that are well-drained. Sandy loams are good as they are low in sulphur, while clayey soils usually have a high sulphur content and produce pungent bulbs. Onions require a high level of nutrients in the soil. Phosphorus is often present in sufficient quantities, but may be applied before planting because of its low level of availability in cold soils. Nitrogen and potash can be applied at regular intervals during the growing season, the last application of nitrogen being at least four weeks before harvesting. Bulbing onions are day-length sensitive; their bulbs begin growing only after the number of daylight hours has surpassed some minimal quantity. Most traditional European onions are referred to as "long-day" onions, producing bulbs only after 14 hours or more of daylight occurs. Southern European and North African varieties are often known as "intermediate-day" types, requiring only 12–13 hours of daylight to stimulate bulb formation. Finally, "short-day" onions, which have been developed in more recent times, are planted in mild-winter areas in the fall and form bulbs in the early spring, and require only 11–12 hours of daylight to stimulate bulb formation. Onions are a cool-weather crop and can be grown in USDA zones 3 to 9. Hot temperatures or other stressful conditions cause them to "bolt", meaning that a flower stem begins to grow.

Large-scale cultivation

Onions may be grown from seed or from sets. Onion seeds are short-lived and fresh seeds germinate better. The seeds are sown thinly in shallow drills, thinning the plants in stages. In suitable climates, certain cultivars can be sown in late summer and autumn to overwinter in the ground and produce early crops the following year. Onion sets are produced by sowing seed thickly in early summer in poor soil and the small bulbs produced are harvested in the autumn. These bulbs are planted the following spring and grow into mature bulbs later in the year. Certain cultivars are used for this purpose and these may not have such good storage characteristics as those grown directly from seed.

Routine care during the growing season involves keeping the rows free of competing weeds, especially when the plants are young. The plants are shallow-rooted and do not need a great deal of water when established. Bulbing usually takes place after 12 to 18 weeks. The bulbs can be gathered when needed to eat fresh, but if they will be kept in storage, they should be harvested after the leaves have died back naturally. In dry weather, they can be left on the surface of the soil for a few days to dry out properly, then they can be placed in nets, roped into strings, or laid in layers in shallow boxes. They should be stored in a well-ventilated, cool place such as a shed.

Pests and Diseases

Onions suffer from a number of plant disorders. The most serious for the home gardener are likely to be the onion fly, stem and bulb eelworm, white rot, and neck rot. Diseases affecting the foliage include rust and smut, downy mildew, and white tip disease. The bulbs may be affected by splitting, white rot, and neck rot. Shanking is a condition in which the central leaves turn yellow and the inner part of the bulb collapses into an unpleasant-smelling slime. Most of these disorders are best treated by removing and burning affected plants. The larvae of the onion leaf miner or leek moth (*Acrolepiopsis assectella*) sometimes attack the foliage and may burrow down into the bulb.

Larvae of the onion fly

The onion fly (*Delia antiqua*) lays eggs on the leaves and stems and on the ground close to onion, shallot, leek, and garlic plants. The fly is attracted to the crop by the smell of damaged tissue and is liable to occur after thinning. Plants grown from sets are less prone to attack. The larvae tunnel into the bulbs and the foliage wilts and turns yellow. The bulbs are disfigured and rot, especially in wet weather. Control measures may include crop rotation, the use of seed dressings, early sowing or planting, and the removal of infested plants.

The onion eelworm (*Ditylenchus dipsaci*), a tiny parasitic soil-living nematode, causes swollen, distorted foliage. Young plants are killed and older ones produce soft bulbs. No cure is known and affected plants should be uprooted and burned. The site should not be used for growing onions again for several years and should also be avoided for growing carrots, parsnips, and beans, which are also susceptible to the eelworm.

White rot of onions, leeks, and garlic is caused by the soil-borne fungus *Sclerotium cepivorum*. As the roots rot, the foliage turns yellow and wilts. The bases of the bulbs are attacked and become covered by a fluffy white mass of mycelia, which later produces small, globular black structures called sclerotia. These resting structures remain in the soil to reinfect a future crop. No cure for this fungal disease exists, so affected plants should be removed and destroyed and the ground used for unrelated crops in subsequent years.

Neck rot is a fungal disease affecting onions in storage. It is caused by *Botrytis allii*, which attacks the neck and upper parts of the bulb, causing a grey mould to develop. The symptoms often first occur where the bulb has been damaged and spread downwards in the affected scales. Large quantities of spores are produced and crust-like sclerotia may also develop. In time, a dry rot sets in and the bulb becomes a dry, mummified structure. This disease may be present throughout the growing period, but only manifests itself when the bulb is in storage. Antifungal seed dressings are available and the disease can be minimized by preventing physical damage to the bulbs at harvesting, careful drying and curing of the mature onions, and correct storage in a cool, dry place with plenty of circulating air.

Storage in the Home

Cooking onions and sweet onions are better stored at room temperature, optimally in a single layer, in mesh bags in a dry, cool, dark, well-ventilated location. In this environment, cooking onions have a shelf life of three to four weeks and sweet onions one to two weeks. Cooking onions will absorb odors from apples and pears. Also, they draw moisture from vegetables with which they are stored which may cause them to decay.

Onion at end of storage life, beginning to sprout

Sweet onions have a greater water and sugar content than cooking onions. This makes them sweeter and milder tasting, but reduces their shelf life. Sweet onions can be stored refrigerated; they have a shelf life of around 1 month. Irrespective of type, any cut pieces of onion are best tightly wrapped, stored away from other produce, and used within two to three days.

Varieties

Common Onion Group (var. cepa)

Most of the diversity within *A. cepa* occurs within this group, the most economically important *Allium* crop. Plants within this group form large single bulbs, and are grown from seed or seed-grown sets. The majority of cultivars grown for dry bulbs, salad onions, and pickling onions belong to this group.

Rossa di Tropea onions for sale in Italy

The range of diversity found among these cultivars includes variation in photoperiod (length of day that triggers bulbing), storage life, flavour, and skin colour. Common onions range from the pungent varieties used for dried soups and onion powder to the mild and hearty sweet onions, such as the Vidalia from Georgia, USA, or Walla Walla from Washington that can be sliced and eaten raw on a sandwich.

Aggregatum Group (var. aggregatum)

This group contains shallots and potato onions, also referred to as multiplier onions. The bulbs are smaller than those of common onions, and a single plant forms an aggregate cluster of several bulbs. They are propagated almost exclusively from daughter bulbs, although reproduction from seed is possible. Shallots are the most important subgroup within this group and comprise the only cultivars cultivated commercially. They form aggregate clusters of small, narrowly ovoid to pear-shaped bulbs. Potato onions differ from shallots in forming larger bulbs with fewer bulbs per cluster, and having a flattened (onion-like) shape. However, intermediate forms exist.

I'itoi onion is a prolific multiplier onion cultivated in the Baboquivari Peak Wilderness, Arizona area. This small-bulb type has a shallot-like flavor and is easy to grow and ideal for hot, dry climates. Bulbs are separated, and planted in the fall 1 in below the surface and 12 in apart. Bulbs will multiply into clumps and can be harvested throughout the cooler months. Tops die back in the heat of summer and may return with heavy rains; bulbs can remain in the ground or be harvested and stored in a cool dry place for planting in the fall. The plants rarely flower; propagation is by division.

Hybrids with A. cepa parentage

A number of hybrids are cultivated that have *A. cepa* parentage, such as the diploid tree onion or Egyptian onion (*A.* ×*proliferum*), Wakegi onion (*A.* ×*wakegi*), and the triploid onion (*A.* ×*cornutum*).

A. ×*proliferum*, tree onion

The tree onion or Egyptian onion produces bulblets in the umbel instead of flowers, and is now known to be a hybrid of *A. cepa* × *A. fistulosum*. It has previously been treated as a variety of *A. cepa*, for example *A. cepa* var. *proliferum*, *A. cepa* var. *bulbiferum*, and *A. cepa* var. *viviparum*.

The Wakegi onion is also known to be a hybrid between *A. cepa* and *A. fistulosum*, with the *A. cepa* parent believed to be from the Aggregatum group of cultivars. It has been grown for centuries in Japan and China for use as a salad onion.

Under the rules of botanical nomenclature, both the Egyptian onion and Wakegi onion should be combined into one hybrid species, having the same parent species. Where this is followed, the Egyptian onion is named *A.* ×*proliferum* Eurasian group and the Wakegi onion is named *A.* ×*proliferum* East Asian group.

The triploid onion is a hybrid species with three sets of chromosomes, two sets from *A. cepa* and the third set from an unknown parent. Various clones of the triploid onion are grown locally in different regions, such as 'Ljutika' in Croatia, and 'Pran', 'Poonch', and 'Srinagar' in the India-Kashmir region. 'Pran' is grown extensively in the northern Indian provinces of Jammu and Kashmir. There are very small genetic differences between 'Pran' and the Croatian clone 'Ljutika', implying a monophyletic origin for this species.

Some authors have used the name *A. cepa* var. *viviparum* (Metzg.) Alef. for the triploid onion, but this name has also been applied to the Egyptian onion. The only name unambiguously connected with the triploid onion is *A.* ×*cornutum*.

Spring onions or salad onions may be grown from the Welsh onion (*A. fistulosum*), as well as from *A. cepa*. Young plants of *A. fistulosum* and *A. cepa* look very similar, but may be distinguished by their leaves, which are circular in cross-section in *A. fistulosum* rather than flattened on one side.

Production and Trade

An estimated 9,000,000 acres (3,642,000 ha) of onions are grown around the world, annually. About 170 countries cultivate onions for domestic use and about 8% of the global production is traded internationally.

Top Ten Onions (dry) Producers — 2012 (metric tons)	
China	20,507,759
India	13,372,100
United States	3,320,870
Egypt	2,208,080
Iran	1,922,970
Turkey	1,900,000
Pakistan	1,701,100
Brazil	1,556,000

Russia	1,536,300
Republic of Korea	1,411,650
World Total	74,250,809
Source: UN Food & Agriculture Organisation (FAO)	

The Onion Futures Act, passed in 1958, bans the trading of futures contracts on onions in the United States. This prohibition came into force after farmers complained about alleged market manipulation by Sam Seigel and Vincent Kosuga at the Chicago Mercantile Exchange two years earlier. The subsequent investigation provided economists with a unique case study into the effects of futures trading on agricultural prices. The act remains in effect as of 2013.

GARLIC

Allium sativum, commonly known as garlic, is a species in the onion genus, *Allium*.

Its close relatives include the onion, shallot, leek, chive, and rakkyo. With a history of over 7,000 years of human consumption and use, garlic is native to central Asia, and has long been a staple in the Mediterranean region, as well as a frequent seasoning in Asia, Africa, and Europe. It was known to Ancient Egyptians, and has been used both as a food flavoring and as a traditional medicine.

Description

Allium sativum is a bulbous plant. It grows up to 1.2 m (4 ft) in height. Its hardiness is USDA Zone 8. It produces hermaphrodite flowers. It is pollinated by bees and other insects.

Origin and Major Types

Allium sativum grows in the wild in areas where it has become naturalized. The "wild garlic", "crow garlic", and "field garlic" of Britain are members of the species *Allium ursinum*, *Allium vineale*, and *Allium oleraceum*, respectively. Identification of the wild progenitor of common garlic is difficult, due to the sterility of its many cultivars which may all be descended from the species *Allium longicuspis*, which grows wild in central and southwestern Asia.

In North America, *Allium vineale* (known as "wild garlic" or "crow garlic") and *Allium canadense*, known as "meadow garlic" or "wild garlic" and "wild onion", are common weeds in fields. So-called elephant garlic is actually a wild leek (*Allium ampeloprasum*), and not a true garlic. Single clove garlic (also called pearl or solo garlic) originated in the Yunnan province of China.

European Garlic

Flower head

There are a number of garlics with Protected Geographical Status in Europe; these include:

Name	Source
Aglio Rosso di Nubia (Red Garlic of Nubia)	Nubia-Paceco, Provincia di Trapani, Sicily, Italy
Aglio Bianco Polesano	Veneto, Italy (PDO)
Aglio di Voghiera	Ferrara, Emilia-Romagna, Italy (PDO)
Ail blanc de Lomagne	Lomagne in the Gascony, France (PGI)
Ail de la Drôme	Drôme, France (PGI)
Ail rose de Lautrec, a rose/pink garlic	Lautrec, France (PGI)
Ajo Morado de las Pedroñeras, a rose/pink garlic	Las Pedroñeras, Spain (PGI)

Italian garlic

Bulbils

Subspecies and Varieties

There are two subspecies of *A. sativum,* ten major groups of varieties, and hundreds of varieties or cultivars.

- *A. sativum* var. *ophioscorodon* (Link) Döll, called *Ophioscorodon*, or hard-necked garlic, includes porcelain garlics, rocambole garlic, and purple stripe garlics. It is sometimes considered to be a separate species, *Allium ophioscorodon* G.Don.
- *A. sativum* var. *sativum*, or soft-necked garlic, includes artichoke garlic, silverskin garlic, and creole garlic.

Cultivation

Garlic is easy to grow and can be grown year-round in mild climates. While sexual propagation of garlic is possible, nearly all of the garlic in cultivation is propagated asexually, by planting individual cloves in the ground. In colder climates, cloves are planted in the autumn, about six weeks before the soil freezes, and harvested in late spring or early summer. The cloves must be planted deep enough to prevent freeze/thaw, which causes mold or white rot. Garlic plants are usually very hardy, and are not attacked by many pests or diseases. Garlic plants are said to repel rabbits and moles. However, two of the major pathogens that attack garlic are nematodes and white rot disease, which remain in the soil indefinitely after the ground has become infected. Garlic can also suffer from pink root, a typically nonfatal disease that stunts the roots and turns them pink or red.

Garlic plants can be grown closely together, leaving enough space for the bulbs to mature, and are easily grown in containers of sufficient depth. Garlic does well in loose, dry, well-drained soils in sunny locations, and is hardy throughout USDA climate zones 4–9. When selecting garlic for planting, it is important to pick large bulbs from which to separate cloves. Large cloves, along with proper spacing in the planting bed, will also improve bulb size. Garlic plants prefer to grow in a soil with a high organic material content, but are capable of growing in a wide range of soil conditions and pH levels.

There are different varieties or subspecies of garlic, most notably hardneck garlic and softneck garlic. The latitude where the garlic is grown affects the choice of type, as garlic can be day-length sensitive. Hardneck garlic is generally grown in cooler climates; softneck garlic is generally grown closer to the equator.

Garlic scapes are removed to focus all the garlic's energy into bulb growth. The scapes can be eaten raw or cooked.

Production Trends

Garlic is grown globally, but China is by far the largest producer of garlic, with around 19 million tonnes (42 billion pounds) grown annually, accounting for over 79% of world output. India (5.2%) and South Korea (1.7%) follow, with Egypt (0.9%) in fourth place and the United States (where garlic is grown in every state except for Alaska) in tenth place (0.7%). This leaves 11% of global garlic production in countries that each produce less than 2% of global output. Much of the garlic production in the United States is centered in Gilroy, California, which calls itself the "garlic capital of the world".

Top 10 garlic producers in 2013

Country	**Production (tonnes)**	**Footnote**		
China	19,168,800			
India	1,259,000			
South Korea	412,250			
Egypt	234,164			
Russia	232,843			
Bangladesh	224,000			
Myanmar	212,000	F		
Uzbekistan	203,585			
Ukraine	185,600			
United States	175,445			
World	**24,285,303**	**A**		

* = Unofficial figure | [ ] = Official data | A = May include official, semi-official or estimated data
F = FAO estimate | Im = FAO data based on imputation methodology | M = Data not available
Source: UN Food & Agriculture Organisation (FAO)

LEEK

The leek is a vegetable, a cultivar of *Allium ampeloprasum*, the broadleaf wild leek. The edible part of the plant is a bundle of leaf sheaths that is sometimes erroneously called a stem or stalk. Historically, many scientific names were used for leeks, but they are now all treated as cultivars of *A. ampeloprasum*. The name 'leek' developed from the Anglo-Saxon word *leac*. Two closely related vegetables, elephant garlic and kurrat, are also cultivars of *A. ampeloprasum*, although different in their uses as food. The onion and garlic are also related, being other species of the genus *Allium*.

Form

Rather than forming a tight bulb like the onion, the leek produces a long cylinder of bundled leaf sheaths that are generally blanched by pushing soil around them (trenching). They are often sold as small seedlings in flats that are started off early in greenhouses, to be planted out as weather permits. Once established in the garden, leeks are hardy; many varieties can be left in the ground during the winter to be harvested as needed.

Cultivars

Leek cultivars may be treated as a single cultivar group, e.g. as *A. ampeloprasum* 'Leek Group'. The cultivars can be subdivided in several ways, but the most common types are "summer leeks", intended for harvest in the season when planted, and overwintering leeks, meant to be harvested in the spring of the year following planting. Summer leek types are generally smaller than overwintering types; overwintering types are generally more strongly flavored. Cultivars include 'King Richard' and 'Tadorna Blue'.

Growing

Leeks are easy to grow from seed and tolerate standing in the field for an extended harvest, which takes place up to 6 months from planting. The soil in which it is grown has to be loose and drained well; leek can be grown in the same regions where onions can be grown. Leeks usually reach maturity in the autumn months. The thrips species *Thrips tabaci* is considered a leek pest and leeks can also get leek rust (*Puccinia allii*). Leeks can be bunched and harvested early when they are about the size of a finger or pencil, or they can be thinned and allowed to grow to a much larger mature size. Hilling leeks can produce better specimens.

SHALLOT

The shallot is a type of onion, specifically a botanical variety of the species *Allium cepa*.

The shallot was formerly classified as a separate species, *A. ascalonicum*, a name now considered a synonym of the currently accepted name.

Shallots, a smaller onion variant used for cooking. These were bought in the American Northeast.

Names

Shallots probably originated in Central or Southwest Asia, travelling from there to India and the eastern Mediterranean. The name "shallot" comes from Ashkelon, an ancient Canaanite city, where people in classical Greek times believed shallots originated.

Sliced and whole shallots

Shallots are called "small onions" in South India and are used extensively in cooking there.

Indian names for shallots include *kaanda* or *gandana* or *pyaaz* (Hindi, Marathi, Marwari and Punjabi), *gundhun* (Bengali), *cheriya ulli* or *chuvanna ulli* (Malayalam), ulli piaja in Odia, *chinna ullipayi* (Telugu) and *chinna vengayam* (or *sambar vengayam* in the Chennai region) (Tamil). In the Kashmiri language, shallots are called *praan*. In Nepal, shallots are called *chyapi* .

In Southeastern Asia, shallots are called *bawang merah kecil* (small red onions) in Malay, *brambang* in Java, *sibuyas bombay* (Indian onion) in the Philippines (in contrast with *sibuyas Tagalog*, the larger, red onion), and *hom* (หอม, fragrant) in Thai. In Cambodian

(Khmer), shallots are called *katem kror hom*, where *katem* or *ktem* is a species of onion, and *kror hom* or *hom* meaning "red", describes their colour.

The name *shallot* is also used for the Persian shallot *(A. stipitatum)*, from the Zagros Mountains in Iran and Iraq. The term *shallot* is further used for the French red shallot (*Allium cepa* var. *aggregatum*, or the *A. cepa* Aggregatum Group) and the French gray shallot or griselle (Allium oschaninii), a species referred to as "true shallot"; it grows wild from Central to Southwest Asia. The name *shallot* is also used for a scallion in New Orleans while the term *French shallot* refers to the plant referred to on this page.

The term *eschalot*, derived from the French word *échalote*, can also be used to refer to the shallot. The usage of *green onion* for shallot is found among English-speaking people in Quebec; but when *shallot* is used, stress is on the second syllable.

Description and Cultivation

Shallot plant (*A. cepa var. aggregatum*) growing in Castelltallat, Spain

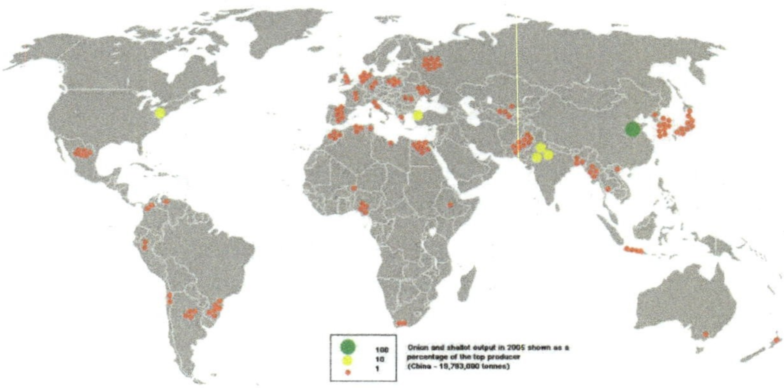

Onion and shallot output in 2005

Whole shallot plant, consist of roots, bulbs, leaves, stalks, and flowers

Like garlic, shallots are formed in clusters of offsets with a head composed of multiple cloves. The skin colour of shallots can vary from golden brown to gray to rose red, and their off-white flesh is usually tinged with green or magenta.

Shallots on sale in India

Shallots are extensively cultivated for culinary uses, propagated by offsets. In some regions ("long-season areas"), the offsets are usually planted in autumn (September or October in the Northern Hemisphere). In some other regions, the suggested planting time for the principal crop is early spring (typically in February or the beginning of March in the Northern Hemisphere).

In planting, the tops of the bulbs should be kept a little above ground, and the soil surrounding the bulbs is often drawn away when the roots have taken hold. They come to maturity in summer, although fresh shallots can now be found year-round in supermarkets. Shallots should not be planted on ground recently manured.

In Africa, shallots are grown in the area around Anloga in southeastern Ghana.

CHIVES

Chives is the common name of *Allium schoenoprasum*, an edible species of the *Allium* genus.

A perennial plant, it is widespread in nature across much of Europe, Asia, and North America.

A. schoenoprasum is the only species of *Allium* native to both the New and the Old Worlds.

The name of the species derives from the Greek *skhoínos* (sedge) and *práson* (leek). Its English name, chives, derives from the French word *cive*, from *cepa*, the Latin word for onion.

Chives are a commonly used herb and can be found in grocery stores or grown in home gardens. In culinary use, the scapes and the unopened, immature flower buds are diced and used as an ingredient for fish, potatoes, soups, and other dishes. Chives have insect-repelling properties that can be used in gardens to control pests.

Allium schoenoprasum, in a garden in Belgium.

Allium schoenoprasum L. Original Description Gräslök, *Allium schoenoprasum* L.

Biology

Chives are a bulb-forming herbaceous perennial plant, growing to 30–50 cm (12–20 in) tall. The bulbs are slender, conical, 2–3 cm ($\frac{3}{4}$–1 $\frac{1}{4}$ in) long and 1 cm ($\frac{1}{2}$ in) broad, and grow in dense clusters from the roots. The scapes (or stems) are hollow and tubular, up to 50 cm (20 in) long and 2–3 mm ($\frac{1}{16}$–$\frac{1}{8}$ in) across, with a soft texture, although, prior to the emergence of a flower, they may appear stiffer than usual. The leaves, which are shorter than the scapes, are also hollow and tubular, or terete, (round in cross-section) which distinguishes it at a glance from garlic chives (*Allium tuberosum*). The flowers are pale purple, and star-shaped with six petals, 1–2 cm ($\frac{1}{2}$–$\frac{3}{4}$ in) wide, and produced in a dense inflorescence of 10-30 together; before opening, the inflorescence is surrounded by a papery bract. The seeds are produced in a small, three-valved capsule, maturing in summer. The herb flowers from April to May in the southern parts of its habitat zones and in June in the northern parts.

Chives are the only species of *Allium* native to both the New and the Old Worlds. Sometimes, the plants found in North America are classified as *A. schoenoprasum* var. *sibiricum*, although this is disputed. Differences between specimens are significant. One example was found in northern Maine growing solitary, instead of in clumps, also exhibiting dingy grey flowers.

Although chives are repulsive to insects in general, due to their sulfur compounds, their flowers attract bees, and they are at times kept to increase desired insect life.

Cultivation

Chives are cultivated both for their culinary uses and their ornamental value; the violet flowers are often used in ornamental dry bouquets.

Chives thrive in well-drained soil, rich in organic matter, with a pH of 6-7 and full sun. They can be grown from seed and mature in summer, or early the following spring. Typically, chives need to be germinated at a temperature of 15 to 20 °C (60-70 °F) and kept moist. They can also be planted under a cloche or germinated indoors in cooler climates, then planted out later. After at least four weeks, the young shoots should be ready to be planted out. They are also easily propagated by division.

Flower with bud

Fully open flower

In cold regions, chives die back to the underground bulbs in winter, with the new leaves appearing in early spring. Chives starting to look old can be cut back to about 2–5 cm. When harvesting, the needed number of stalks should be cut to the base. During the growing season, the plant continually regrows leaves, allowing for a continuous harvest.

History and Cultural Importance

Chives have been cultivated in Europe since the Middle Ages (fifth until the 15th centuries), although their usage dates back 5000 years. They were sometimes referred to as "rush leeks" (from the Greek *schoinos* meaning rush and *prason* meaning leek).

The Romans believed chives could relieve the pain from sunburn or a sore throat. They believed eating chives could increase blood pressure and act as a diuretic.

Romanian Gypsies have used chives in fortune telling. Bunches of dried chives hung around a house were believed to ward off disease and evil.

REFERENCES

1. 1793 illustration from William Woodville: "Medical botany", London, James Phillips, 1793, Vol. 3, Plate 168: Allium sativum (Garlic). Hand-coloured engraving
2. Linnaeus 1753, Allium I pp. 294–301
3. Kew World Checklist of Selected Plant Families
4. Quattrocchi 1999, vol. 1 p. 91.
5. Eric Block (2010). Garlic and Other Alliums: The Lore and the Science. Royal Society of Chemistry. ISBN 978-0-85404-190-9.
6. Dilys Davies (1992). Alliums: The Ornamental Onions. Timber Press. ISBN 0-88192-241-2.
7. Knud Rahn. 1998. "Alliaceae" pages 70-78. In: Klaus Kubitzki (editor). The Families and Genera of Vascular Plants volume III. Springer-Verlag: Berlin;Heidelberg, Germany. ISBN 978-3-540-64060-8
8. The Plant List, for genus Allium
9. Allium In: Index Nominum Genericorum. In: Regnum Vegetabile.
10. Chase, M.W.; Reveal, J.L. & Fay, M.F. (2009), "A subfamilial classification for the expanded asparagalean families Amaryllidaceae, Asparagaceae and Xanthorrhoeaceae", Botanical Journal of the Linnean Society, 161 (2): 132–136, doi:10.1111/j.1095-8339.2009.00999.x
11. James L. Brewster, "Onions and Other Alliums" (Wallingford: CABI Publishing, 2008)
12. Haim D. Rabinowitch, Leslie Currah, "Allium Crop Sciences: Recent Advances" (Wallingford: CABI Publishing, 2002)
13. Penny Woodward, "Garlic and Friends: The History, Growth and Use of Edible Alliums" (South Melbourne: Hyland House, 1996)

14. Frodin, David G. (2004). "History and concepts of big plant genera". Taxon. 53 (3): 753–776. doi:10.2307/4135449.
15. "Allium in Flora of China @". Efloras.org. Retrieved 2012-12-11.
16. "Allium in Flora of North America @". Efloras.org. Retrieved 2012-12-11.
17. Peterson, R.P. 1982. A Field Guide to Edible Wild Plants: Eastern and central North America. Houghton Mifflin, Boston.
18. Gibbons, E. 1962. Stalking the wild asparagus. David McKay, New York.
19. Deniz et al 2015.
20. Friesen et al 2000.
21. Linnaeus Sexual System 2015.
22. Linnaeus 1753, Hexandria monogynia I pp. 285–332.
23. von Berg et al 1996.
24. Friesen, Fritsch & Blattner 2006.
25. Sykorova 2006.
26. Fritsch et al 2010.
27. Nguyen et al 2008.
28. Hirschegger et al 2010.
29. Gualtiero Simonetti (1990). Stanley Schuler, ed. Simon & Schuster's Guide to Herbs and Spices. Simon & Schuster, Inc. ISBN 0-671-73489-X.
30. Anthony Huxley, Mark Griffiths, and Margot Levy (1992). The New Royal Horticultural Society Dictionary of Gardening. The Macmillan Press,Limited: London. The Stockton Press: New York. ISBN 978-0-333-47494-5 (set).
31. Brickell, Christopher (Editor-in-chief),The Royal Horticultural Society A–Z Encyclopedia of Garden Plants, p.95, Dorling Kindersley, London, 1996, ISBN 0-7513-0303-8
32. RHS Plant Finder 2009–2010, p68, Dorling Kindersley, London, 2009, ISBN 978-1-4053-4176-9
33. Lloyd, Christopher & Rice, Graham, (1991) Garden Flowers From Seed, p45, Viking, ISBN 0-670-82455-0
34. "RHS Plant Selector Allium 'Beau Regard' AGM / RHS Gardening". Apps.rhs.org.uk. Retrieved 2012-12-11.
35. "RHS Plant Selector Allium 'Gladiator' AGM / RHS Gardening". Apps.rhs.org.uk. Retrieved 2012-12-11.
36. "RHS Plant Selector Allium 'Globemaster' AGM / RHS Gardening". Apps.rhs.org.uk. Retrieved 2012-12-11.
37. Cope RB. Toxicology Brief: Allium species poisoning in dogs and cats. Veterinary Medicine 2005

CHAPTER 5

Fruit Farmings: An overview

Chapter Contents

- **Introduction**
- **Fruit Structure**
- **Fruit Development**
- **Seedless Fruits**
- **Accessory Fruit**
- **Compound Fruit**

Fruit is a seed-bearing part of a flowering plant or tree that can be eaten as food. By those standards, foods such as avocados, cucumbers, squash, and yes, even tomatoes are all fruits. From a culinary viewpoint, a fruit is usually thought of as any sweet-tasting plant product with seeds, whereas a vegetable is any savory or less sweet-tasting plant.

INTRODUCTION

In botany, a fruit is the seed-bearing structure in flowering plants (also known as angiosperms) formed from the ovary after flowering.

Fruits are the means by which angiosperms disseminate seeds. Edible fruits, in particular, have propagated with the movements of humans and animals in a symbiotic relationship as a means for seed dispersal and nutrition; in fact, humans and many animals have become dependent on fruits as a source of food. Accordingly, fruits account for a substantial fraction of the world's agricultural output, and some (such as the apple and the pomegranate) have acquired extensive cultural and symbolic meanings.

In common language usage, "fruit" normally means the fleshy seed-associated structures of a plant that are sweet or sour, and edible in the raw state, such as apples, bananas, grapes, lemons, oranges, and strawberries. On the other hand, in botanical usage, "fruit" includes many structures that are not commonly called "fruits", such as bean pods, corn kernels, tomatoes, and wheat grains. The section of a fungus that produces spores is also called a fruiting body.

Botanic Fruit and Culinary Fruit

Many common terms for seeds and fruit do not correspond to the botanical classifications. In culinary terminology, a *fruit* is usually any sweet-tasting plant part, especially a botanical fruit; a *nut* is any hard, oily, and shelled plant product; and a *vegetable* is any savory or less sweet plant product. However, in botany, a *fruit* is the ripened ovary or carpel that contains seeds, a *nut* is a type of fruit and not a seed, and a *seed* is a ripened ovule.

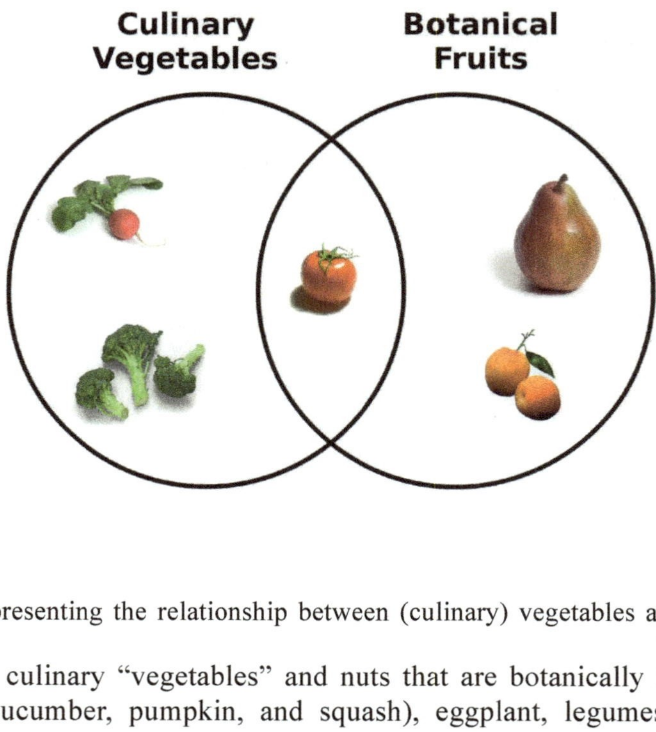

Venn diagram representing the relationship between (culinary) vegetables and botanical fruits

Examples of culinary "vegetables" and nuts that are botanically fruit include corn, cucurbits (e.g., cucumber, pumpkin, and squash), eggplant, legumes (beans, peanuts, and peas), sweet pepper, and tomato. In addition, some spices, such as allspice and chili pepper, are fruits, botanically speaking. In contrast, rhubarb is often referred to as a fruit, because it is used to make sweet desserts such as pies, though only the petiole (leaf stalk) of the rhubarb plant is edible, and edible gymnosperm seeds are often given fruit names, e.g., ginkgo nuts and pine nuts.

Botanically, a cereal grain, such as corn, rice, or wheat, is also a kind of fruit, termed a caryopsis. However, the fruit wall is very thin and is fused to the seed coat, so almost all of the edible grain is actually a seed.

FRUIT STRUCTURE

The outer, often edible layer, is the *pericarp*, formed from the ovary and surrounding the seeds, although in some species other tissues contribute to or form the edible portion. The pericarp may be described in three layers from outer to inner, the *epicarp*, *mesocarp* and *endocarp*.

Fruit that bears a prominent pointed terminal projection is said to be *beaked*.

FRUIT DEVELOPMENT

A fruit results from maturation of one or more flowers, and the gynoecium of the flower(s) forms all or part of the fruit.

The development sequence of a typical drupe, the nectarine (*Prunus persica*) over a 7.5 month period, from bud formation in early winter to fruit ripening in midsummer

Inside the ovary/ovaries are one or more ovules where the megagametophyte contains the egg cell. After double fertilization, these ovules will become seeds. The ovules are fertilized in a process that starts with pollination, which involves the movement of pollen from the stamens to the stigma of flowers. After pollination, a tube grows from the pollen through the stigma into the ovary to the ovule and two sperm are transferred from the pollen to the megagametophyte. Within the megagametophyte one of the two sperm unites with the egg, forming a zygote, and the second sperm enters the central cell forming the endosperm mother cell, which completes the double fertilization process. Later the zygote will give rise to the embryo of the seed, and the endosperm mother cell will give rise to endosperm, a nutritive tissue used by the embryo.

As the ovules develop into seeds, the ovary begins to ripen and the ovary wall, the *pericarp*, may become fleshy (as in berries or drupes), or form a hard outer covering (as in nuts). In some multiseeded fruits, the extent to which the flesh develops is proportional to the number of fertilized ovules. The pericarp is often differentiated into two or three distinct layers called the *exocarp* (outer layer, also called epicarp), *mesocarp* (middle layer), and *endocarp* (inner layer). In some fruits, especially simple fruits derived from an inferior ovary, other parts of the flower (such as the floral tube, including the petals, sepals, and stamens), fuse with the ovary and ripen with it. In other cases, the sepals, petals and/or stamens and style of the flower fall off. When such other floral parts are a significant part of the fruit, it is called an *accessory fruit*. Since other parts of the flower may contribute to the structure of the fruit, it is important to study flower structure to understand how a particular fruit forms.

There are three general modes of fruit development:

- Apocarpous fruits develop from a single flower having one or more separate carpels, and they are the simplest fruits.
- Syncarpous fruits develop from a single gynoecium having two or more carpels fused together.
- Multiple fruits form from many different flowers.

Plant scientists have grouped fruits into three main groups, simple fruits, aggregate fruits, and composite or multiple fruits. The groupings are not evolutionarily relevant, since many diverse plant taxa may be in the same group, but reflect how the flower organs are arranged and how the fruits develop.

Simple Fruit

Simple fruits can be either dry or fleshy, and result from the ripening of a simple or compound ovary in a flower with only one pistil.

Epigynous berries are simple fleshy fruit. Clockwise from top right: cranberries, lingonberries, blueberries, red huckleberries

Dry fruits may be either dehiscent (they open to discharge seeds), or indehiscent (they do not open to discharge seeds). Types of dry, simple fruits, and examples of each, include:

- achene – Most commonly seen in aggregate fruits (e.g., strawberry)
- capsule – (e.g., Brazil nut)
- caryopsis – (e.g., wheat)
- cypsela – an achene-like fruit derived from the individual florets in a capitulum (e.g., dandelion).
- fibrous drupe – (e.g., coconut, walnut)
- follicle – is formed from a single carpel, opens by one suture (e.g., milkweed), commonly seen in aggregate fruits (e.g., magnolia)
- legume – (e.g., bean, pea, peanut)
- loment – a type of indehiscent legume
- nut – (e.g., beech, hazelnut, oak acorn)
- samara – (e.g., ash, elm, maple key)
- schizocarp – (e.g., carrot seed)
- silique – (e.g., radish seed)
- silicle – (e.g., shepherd's purse)
- utricle – (e.g., beet)

Lilium unripe capsule fruit

Fruits in which part or all of the *pericarp* (fruit wall) is fleshy at maturity are *simple fleshy fruits*. Types of simple, fleshy, fruits (with examples) include:

- berry – (e.g., cranberry, gooseberry, redcurrant, tomato)
- stone fruit or drupe (e.g., apricot, cherry, olive, peach, plum)

Dewberry flowers. Note the multiple pistils, each of which will produce a drupelet. Each flower will become a blackberry-like aggregate fruit.

An aggregate fruit, or *etaerio*, develops from a single flower with numerous simple pistils.

- Magnolia and peony, collection of follicles developing from one flower.
- Sweet gum, collection of capsules.
- Sycamore, collection of achenes.
- Teasel, collection of cypsellas
- Tuliptree, collection of samaras.

The pome fruits of the family Rosaceae, (including apples, pears, rosehips, and saskatoon berry) are a syncarpous fleshy fruit, a simple fruit, developing from a half-inferior ovary.

Schizocarp fruits form from a syncarpous ovary and do not really dehisce, but rather split into segments with one or more seeds; they include a number of different forms from a wide range of families. Carrot seed is an example.

Aggregate Fruit

Aggregate fruits form from single flowers that have multiple carpels which are not joined together, i.e. each pistil contains one carpel. Each pistil forms a fruitlet, and collectively the fruitlets are called an etaerio. Four types of aggregate fruits include etaerios of achenes, follicles, drupelets, and berries. Ranunculaceae species, including *Clematis* and *Ranunculus* have an etaerio of achenes, *Calotropis* has an etaerio of follicles, and *Rubus* species like raspberry, have an etaerio of drupelets. *Annona* have an etaerio of berries.

Detail of raspberry flower

The raspberry, whose pistils are termed *drupelets* because each is like a small drupe attached to the receptacle. In some bramble fruits (such as blackberry) the receptacle is elongated and part of the ripe fruit, making the blackberry an *aggregate-accessory* fruit. The strawberry is also an aggregate-accessory fruit, only one in which the seeds are contained in achenes. In all these examples, the fruit develops from a single flower with numerous pistils.

Multiple Fruits

A multiple fruit is one formed from a cluster of flowers (called an *inflorescence*). Each flower produces a fruit, but these mature into a single mass. Examples are the pineapple, fig, mulberry, osage-orange, and breadfruit.

In some plants, such as this noni, flowers are produced regularly along the stem and it is possible to see together examples of flowering, fruit development, and fruit ripening.

In the photograph on the right, stages of flowering and fruit development in the noni or Indian mulberry (*Morinda citrifolia*) can be observed on a single branch. First an inflorescence of white flowers called a head is produced. After fertilization, each flower develops into a drupe, and as the drupes expand, they become *connate* (merge) into a *multiple fleshy fruit* called a *syncarp*.

Berries

Berries are another type of fleshy fruit; they are simple fruit created from a single ovary. The ovary may be compound, with several carpels. Types include (examples follow in the table below):

- Pepo – berries whose skin is hardened, cucurbits
- Hesperidium – berries with a rind and a juicy interior, like most citrus fruit

Accessory Fruit

The fruit of a pineapple includes tissue from the sepals as well as the pistils of many flowers. It is an accessory fruit and a multiple fruit.

Some or all of the edible part of accessory fruit is not generated by the ovary. Accessory fruit can be simple, aggregate, or multiple, i.e., they can include one or more pistils and other parts from the same flower, or the pistils and other parts of many flowers.

Table of fruit examples

Types of fleshy fruits					
True berry	**Pepo**	**Hesperidium**	**Aggregate fruit**	**Multiple fruit**	**Accessory fruit**
Blackcurrant, Blueberry, Chili pepper, Cranberry, Eggplant, Gooseberry, Grape, Guava, Kiwifruit, Lucuma, Pomegranate, Redcurrant, Tomato	Cucumber, Gourd, Melon, Pumpkin	Grapefruit, Lemon, Lime, Orange	Blackberry, Boysenberry, Raspberry	Fig, Hedge apple, Mulberry, Pineapple	Apple, Pineapple, Rose hip, Stone fruit, Strawberry

SEEDLESS FRUITS

Seedlessness is an important feature of some fruits of commerce. Commercial cultivars of bananas and pineapples are examples of seedless fruits. Some cultivars of citrus fruits (especially grapefruit, mandarin oranges, navel oranges), satsumas, table grapes, and watermelons are valued for their seedlessness. In some species, seedlessness is the result of *parthenocarpy*, where fruits set without fertilization. Parthenocarpic fruit set may or may not require pollination, but most seedless citrus fruits require a stimulus from pollination to produce fruit.

An arrangement of fruits commonly thought of as vegetables, including tomatoes and various squash

Seedless bananas and grapes are triploids, and seedlessness results from the abortion of the embryonic plant that is produced by fertilization, a phenomenon known as *stenospermocarpy*, which requires normal pollination and fertilization.

Seed Dissemination

Variations in fruit structures largely depend on their seeds' mode of dispersal. This dispersal can be achieved by animals, explosive dehiscence, water, or wind.

Grapes and Mangoes

Some fruits have coats covered with spikes or hooked burrs, either to prevent themselves from being eaten by animals, or to stick to the feathers, hairs, or legs of animals, using them as dispersal agents. Examples include cocklebur and unicorn plant.

The sweet flesh of many fruits is "deliberately" appealing to animals, so that the seeds held within are eaten and "unwittingly" carried away and deposited (i.e., defecated) at a distance from the parent. Likewise, the nutritious, oily kernels of nuts are appealing to rodents (such as squirrels), which hoard them in the soil to avoid starving during the winter, thus giving those seeds that remain uneaten the chance to germinate and grow into a new plant away from their parent.

Other fruits are elongated and flattened out naturally, and so become thin, like wings or helicopter blades, e.g., elm, maple, and tuliptree. This is an evolutionary mechanism to increase dispersal distance away from the parent, via wind. Other wind-dispersed fruit have tiny "parachutes", e.g., dandelion, milkweed, salsify.

Coconut fruits can float thousands of miles in the ocean to spread seeds. Some other fruits that can disperse via water are nipa palm and screw pine.

Some fruits fling seeds substantial distances (up to 100 m in sandbox tree) via explosive dehiscence or other mechanisms, e.g., impatiens and squirting cucumber.

Uses

Many hundreds of fruits, including fleshy fruits (like apple, kiwifruit, mango, peach, pear, and watermelon) are commercially valuable as human food, eaten both fresh and as jams, marmalade and other preserves. Fruits are also used in manufactured foods (e.g., cakes, cookies, ice cream, muffins, or yogurt) or beverages, such as fruit juices (e.g., apple juice, grape juice, or orange juice) or alcoholic beverages (e.g., brandy, fruit beer, or wine).

Fruits are also used for gift giving, e.g., in the form of Fruit Baskets and Fruit Bouquets.

Nectarines are one of many fruits that can be easily stewed.

Oranges, bananas, pears, apples, and a watermelon

Many "vegetables" in culinary *parlance* are botanical fruits, including bell pepper, cucumber, eggplant, green bean, okra, pumpkin, squash, tomato, and zucchini. Olive fruit is pressed for olive oil. Spices like allspice, black pepper, paprika, and vanilla are derived from berries.

Nutritional Value

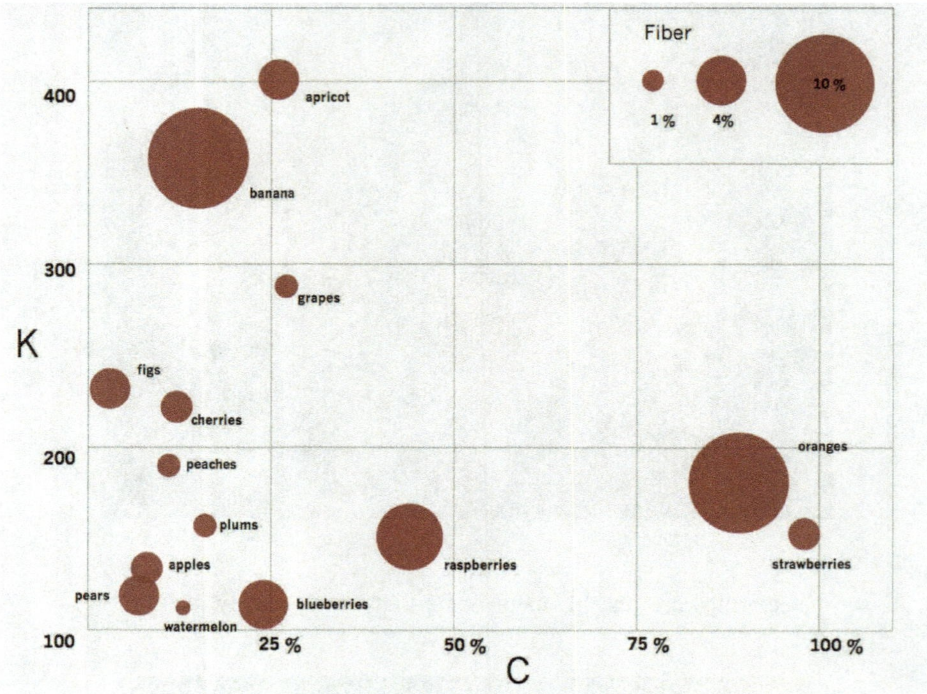

Each point refers to a 100 g serving of the fresh fruit, the daily recommended allowance of vitamin C is on the X axis and mg of Potassium (K) on the Y (offset by 100 mg which every fruit has) and the size of the disk represents amount of fiber (key in upper right). Watermelon, which has almost no fiber, and low levels of vitamin C and potassium, comes in last place.

Fresh fruits are generally high in fiber, vitamin C, and water.

Regular consumption of fruit is generally associated with reduced risks of several diseases and functional declines associated with aging.

Nonfood Uses

Because fruits have been such a major part of the human diet, various cultures have developed many different uses for fruits they do not depend on for food. For example:

- Bayberry fruits provide a wax often used to make candles;
- Many dry fruits are used as decorations or in dried flower arrangements (e.g., annual honesty, cotoneaster, lotus, milkweed, unicorn plant, and wheat). Ornamental trees and shrubs are often cultivated for their colorful fruits, including beautyberry, cotoneaster, holly, pyracantha, skimmia, and viburnum.
- Fruits of opium poppy are the source of opium, which contains the drugs codeine and morphine, as well as the biologically inactive chemical theabaine from which the drug oxycodone is synthesized.

- Osage orange fruits are used to repel cockroaches.
- Many fruits provide natural dyes (e.g., cherry, mulberry, sumac, and walnut).
- Dried gourds are used as bird houses, cups, decorations, dishes, musical instruments, and water jugs.
- Pumpkins are carved into Jack-o'-lanterns for Halloween.
- The spiny fruit of burdock or cocklebur inspired the invention of Velcro.
- Coir fiber from coconut shells is used for brushes, doormats, floor tiles, insulation, mattresses, sacking, and as a growing medium for container plants. The shell of the coconut fruit is used to make bird houses, bowls, cups, musical instruments, and souvenir heads.
- Fruit is often a subject of still life paintings.

Safety

For food safety, the CDC recommends proper fruit handling and preparation to reduce the risk of food contamination and foodborne illness. Fresh fruits and vegetables should be carefully selected; at the store, they should not be damaged or bruised; and precut pieces should be refrigerated or surrounded by ice.

All fruits and vegetables should be rinsed before eating. This recommendation also applies to produce with rinds or skins that are not eaten. It should be done just before preparing or eating to avoid premature spoilage.

Fruits and vegetables should be kept separate from raw foods like meat, poultry, and seafood, as well as from utensils that have come in contact with raw foods. Fruits and vegetables that are not going to be cooked should be thrown away if they have touched raw meat, poultry, seafood, or eggs.

All cut, peeled, or cooked fruits and vegetables should be refrigerated within two hours. After a certain time, harmful bacteria may grow on them and increase the risk of foodborne illness.

Allergies

Fruit allergies make up about 10 percent of all food related allergies.

Storage

All fruits benefit from proper post harvest care, and in many fruits, the plant hormone ethylene causes ripening. Therefore, maintaining most fruits in an efficient cold chain is optimal for post harvest storage, with the aim of extending and ensuring shelf life.

ACCESSORY FRUIT

An accessory fruit (sometimes called *false fruit*, *spurious fruit*, *pseudofruit*, or *pseudocarp*) is a fruit in which some of the flesh is derived not from the ovary but from some adjacent tissue exterior to the carpel.Examples of accessory tissue are the receptacle of the strawberry, pineapple, common fig, and mulberry, and the calyx of *Gaultheria procumbens* or *Syzygium jambos*. Pomes, such as apples and pears, are also accessory fruits, with much of the fruit flesh derived from a hypanthium. Other example could be the *anthocarps* specific to the family Nyctaginaceae, where most of the fruit comes from the perianth (floral whorls).

Fruit with fleshy seeds, such as pomegranate or mamoncillo, are not considered to be accessory fruit.

The terms *false fruit*, *spurious fruit*, and *pseudocarp* are older terms for accessory fruit that have been criticized as "inapt", and are not used by botanists today.

A strawberry fruit: the 'seeds' are achenes, each one derived from a pistil of the flower.

COMPOUND FRUIT

The term compound fruit is not used in technical botanical writing, but is sometimes used when it is not clear which of several fruit types is involved. A compound fruit is "composed of two or more similar parts".

A raspberry is an aggregate fruit (shown with a raspberry beetle larva)

A pineapple is a multiple fruit.

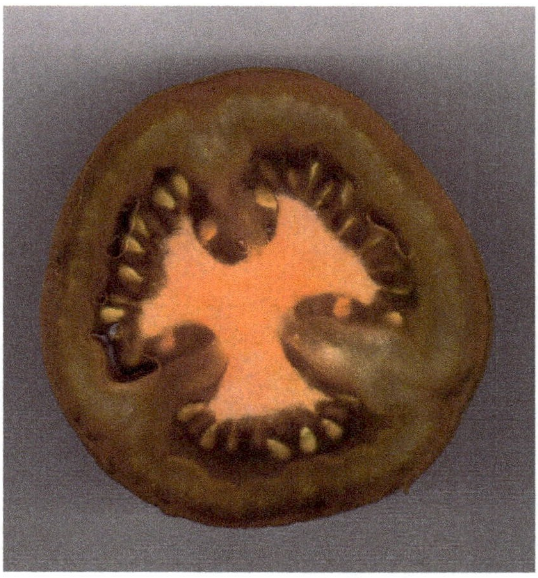

A tomato is a simple fruit derived from a compound ovary.

A compound fruit may be:

- An aggregate fruit, in which one flower contains several separate ovaries, which merge during development.
- A multiple fruit, in which several flowers, each with an ovary, develop into small fruits that are clustered or fused together into a larger fruit.
- A simple fruit formed from a compound ovary.

Grapes grow in clusters, but are not compound fruits. Each grape is grown from one ovary in one flower, and each grape remains an independent fruit.

REFERENCES

1. Lewis, Robert A. (January 1, 2002). CRC Dictionary of Agricultural Sciences. CRC Press. ISBN 0-8493-2327-4.
2. Schlegel, Rolf H J (January 1, 2003). Encyclopedic Dictionary of Plant Breeding and Related Subjects. Haworth Press. p. 177. ISBN 1-56022-950-0.
3. Mauseth, James D. (April 1, 2003). Botany: An Introduction to Plant Biology. Jones and Bartlett. pp. 271–272. ISBN 0-7637-2134-4.
4. "Sporophore from Encyclopædia Britannica".
5. For a Supreme Court of the United States ruling on the matter, see Nix v. Hedden.
6. McGee, Harold (November 16, 2004). On Food and Cooking: The Science and Lore of the Kitchen. Simon & Schuster. pp. 247–248. ISBN 0-684-80001-2.
7. McGee (2004-11-16). On Food and Cooking. p. 367. ISBN 978-0-684-80001-1.
8. Lewis (2002). CRC Dictionary of Agricultural Sciences. p. 238. ISBN 978-0-8493-2327-0.
9. "Glossary of Botanical Terms". FloraBase. Western Australian Herbarium. Retrieved 23 July 2014.
10. Esau, K. 1977. Anatomy of seed plants. John Wiley and Sons, New York.
11. Archived December 20, 2010, at the Wayback Machine.
12. Mauseth, James D. (2003). Botany: an introduction to plant biology. Boston: Jones and Bartlett Publishers. p. 258. ISBN 978-0-7637-2134-3.
13. Rost, Thomas L.; Weier, T. Elliot; Weier, Thomas Elliot (1979). Botany: a brief introduction to plant biology. New York: Wiley. pp. 135–37. ISBN 0-471-02114-8.
14. Mauseth (2003-04-25). Botany. Chapter 9: Flowers and Reproduction. ISBN 978-0-7637-2134-3.
15. Singh, Gurcharan (2004). Plants Systematics: An Integrated Approach. Science Publishers. p. 83. ISBN 1-57808-351-6.
16. Schlegel (2003-05-13). Encyclopedic Dictionary. p. 123. ISBN 978-1-56022-950-6.
17. Schlegel (2003-05-13). Encyclopedic Dictionary. p. 16. ISBN 978-1-56022-950-6.
18. Evolutionary trends in flowering plants. New York: Columbia University Press. 1991. p. 209. ISBN 0-231-07328-3.
19. Gupta, Prof. P.K. Genetics Classical To Modern. Rastogi Publication. pp. 2–134. ISBN 978-81-7133-896-2.
20. http://www.rkv.rgukt.in/content/Biology/47Module/47fruit.pdf[dead link]
21. McGee (2004-11-16). On Food and Cooking. pp. 361–362. ISBN 978-0-684-80001-1.

22. McGee (2004-11-16). On Food and Cooking. pp. 364–365. ISBN 978-0-684-80001-1.
23. Schlegel (2003-05-13). Encyclopedic Dictionary. p. 282. ISBN 978-1-56022-950-6.
24. Spiegel-Roy, P.; E. E. Goldschmidt (August 28, 1996). The Biology of Citrus. Cambridge University Press. pp. 87–88. ISBN 0-521-33321-0.
25. Capon, Brian (February 25, 2005). Botany for Gardeners. Timber Press. pp. 198–199. ISBN 0-88192-655-8.
26. Heiser, Charles B. (April 1, 2003). Weeds in My Garden: Observations on Some Misunderstood Plants. Timber Press. pp. 93–95. ISBN 0-88192-562-4.
27. Heiser (2003-04-01). Weeds in My Garden. pp. 162–164. ISBN 978-0-88192-562-3.
28. Feldkamp, Susan (2002). Modern Biology. Holt, Rinehart, and Winston. p. 634. ISBN 0-88192-562-4.
29. McGee (2004-11-16). On Food and Cooking. Chapter 7: A Survey of Common Fruits. ISBN 978-0-684-80001-1.
30. McGee (2004-11-16). On Food and Cooking. Chapter 6: A Survey of Common Vegetables. ISBN 978-0-684-80001-1.
31. Farrell, Kenneth T. (November 1, 1999). Spices, Condiments and Seasonings. Springer. pp. 17–19. ISBN 0-8342-1337-0.
32. Hulme, A.C (editor) (1970). "The Biochemistry of Fruits and their Products". 1. London & New York: Academic Press.
33. Lim, Stephen S.; Vos, Theo; Flaxman, Abraham D.; Danaei, Goodarz; Shibuya, Kenji; Adair-Rohani, Heather; Amann, Markus; Anderson, H. Ross; Andrews, Kathryn G. (2012-12-15). "A comparative risk assessment of burden of disease and injury attributable to 67 risk factors and risk factor clusters in 21 regions, 1990-2010: a systematic analysis for the Global Burden of Disease Study 2010". Lancet (London, England). 380 (9859): 2224–2260. doi:10.1016/S0140-6736(12)61766-8. ISSN 1474-547X. PMC 4156511Freely accessible. PMID 23245609.
34. Wang X, Ouyang Y, Liu J, Zhu M, Zhao G, Bao W, Hu FB (2014). "Fruit and vegetable consumption and mortality from all causes, cardiovascular disease, and cancer: systematic review and dose-response meta-analysis of prospective cohort studies". BMJ. 349 (Jul 29): g4490. doi:10.1136/bmj.g4490. PMC 4115152Freely accessible. PMID 25073782.
35. K, Amber (December 1, 2001). Candlemas: Feast of Flames. Llewellyn Worldwide. p. 155. ISBN 0-7387-0079-7.
36. Adams, Denise Wiles (February 1, 2004). Restoring American Gardens: An Encyclopedia of Heirloom Ornamental Plants, 1640–1940. Timber Press. ISBN 0-88192-619-1.
37. Booth, Martin (June 12, 1999). Opium: A History. St. Martin's Press. ISBN 0-312-20667-4.

38. Cothran, James R. (November 1, 2003). Gardens and Historic Plants of the Antebellum South. University of South Carolina Press. p. 221. ISBN 1-57003-501-6.
39. Adrosko, Rita J. (June 1, 1971). Natural Dyes and Home Dyeing: A Practical Guide with over 150 Recipes. Courier Dover Publications. ISBN 0-486-22688-3.
40. Wake, Warren (March 13, 2000). Design Paradigms: A Sourcebook for Creative Visualization. John Wiley and Sons. pp. 162–163. ISBN 978-0-471-29976-9.
41. "The Many Uses of the Coconut". The Coconut Museum. Retrieved 2006-09-14.
42. "Nutrition for Everyone: Fruits and Vegetables – DNPAO – CDC". fruitsandveggiesmatter.gov.
43. "Asthma and Allergy Foundation of America". Aafa.org. Retrieved 2014-04-25.
44. Roy Mankovitz (2010). The Wellness Project. Retrieved 2014-04-25.
45. Why Cold Chain for Fruits: Kohli, Pawanexh (2008). "Fruits and Vegetables Post-Harvest Care: The Basics" (PDF). Crosstree Techno-visors.

Cultivation of Aggregate Fruit

Chapter Contents

An aggregate fruit results from the joining together of several ovaries of the same flower. This process starts with a flower that contains several carpels with an equal number of ovaries. Each ovary contains a single ovule that develops into a seed following fertilization. As the ovaries develop and increase in size they become attached together. Aggregate fruits are also either true, without accessory parts, or accessory, with additional floral parts. The raspberry is an aggregate of true fruits. Each fruit in the aggregate is a drupe containing a stony pit. The strawberry is an accessory aggregate fruit. The many ovaries of the strawberry flower develop into achenes which become embedded on the surface of the enlarged flower receptacle.

INTRODUCTION

An aggregate fruit or etaerio is a fruit that develops from the merger of several ovaries that were separate in a single flower. In contrast, a *simple fruit* develops from one ovary. In languages other than English, the meanings of aggregate and multiple fruit are reversed, so that aggregate fruits merge several flowers. The differences in meaning are due to a reversal in the terminology by John Lindley, which has been followed by most English-language authors.

The fruit of an *Aquilegia* flower is one fruit that forms from several ovaries of one flower, and it is an aggregate of follicles. However, because the follicles are not fused to one another, it is not considered an aggregate fruit

Not all flowers with multiple ovaries form aggregate fruit; the ovaries of some flowers do not become tightly joined together to make a larger fruit. Aggregate fruits may also be accessory fruits, in which parts of the flower other than the ovary become fleshy and form part of the fruit.

The individual parts of an aggregate fruit come in many forms. Common examples are:

- drupelets:
  - Raspberry
  - Dewberry and blackberry, also an accessory fruit, with a fleshy receptacle
- achenes:
  - Strawberry, also an accessory fruit, with a fleshy receptacle
  - *Ranunculus*
- follicles:
  - Magnolia
- samaras:
  - Liriodendron tulipifera

A sugar apple fruit forms from the pistils and receptacle of one flower

The components of other aggregate fruit are more difficult to define. For example, sugar apple (*Annona* spp.) fruit are made up of individual berry-like pistils fused with the receptacle.

RASPBERRY

The raspberry is the edible fruit of a multitude of plant species in the genus *Rubus* of the rose family, most of which are in the subgenus *Idaeobatus*; the name also applies to these plants themselves.

Raspberries are perennial with woody stems.

Fruit of four species of raspberry. Clockwise from top left: Boulder raspberry, Korean raspberry, Australian native raspberry, Mauritius raspberry

Cultivation

Various kinds of raspberries can be cultivated from hardiness zones 3 to 9. Raspberries are traditionally planted in the winter as dormant canes, although planting of tender, plug plants produced by tissue culture has become much more common. A specialized production system called "long cane production" involves growing canes for a year in a northern climate such as Scotland or Oregon or Washington, where the chilling requirement for proper bud break is attained, or attained earlier than the ultimate place of planting. These canes are then dug, roots and all, to be replanted in warmer climates such as Spain, where they quickly flower and produce a very early season crop. Plants are typically planted 2-6 per m in fertile, well drained soil; raspberries are usually planted in raised beds/ridges, if there is any question about root rot problems.

Flowering cultivated raspberry

The flowers can be a major nectar source for honeybees and other pollinators.

Raspberries are very vigorous and can be locally invasive. They propagate using basal shoots (also known as suckers), extended underground shoots that develop roots and individual plants. They can sucker new canes some distance from the main plant. For this reason, raspberries spread well, and can take over gardens if left unchecked. Raspberries are often propagated using cuttings, and will root readily in moist soil conditions.

The fruit is harvested when it comes off the receptacle easily and has turned a deep color (red, black, purple, or golden yellow, depending on the species and cultivar). This is when the fruits are ripest and sweetest. Excess fruit can be made into raspberry jam or frozen.

High tunnel bramble production offers the opportunity to bridge gaps in availability during late fall and late spring. Furthermore, high tunnels allow less hardy floricane-fruiting raspberries to overwinter in climates where they wouldn't otherwise survive. In the tunnel plants are established at close spacing usually prior to tunnel construction.

Major Cultivars

Raspberries are an important commercial fruit crop, widely grown in all temperate regions of the world.

Many of the most important modern commercial red raspberry cultivars derive from hybrids between *R. idaeus* and *R. strigosus*. Some botanists consider the Eurasian and American red raspberries to belong to a single, circumboreal species, *Rubus idaeus,* with the European plants then classified as either *R. idaeus* subsp. *idaeus* or *R. idaeus* var. *idaeus,* and the native North American red raspberries classified as either *R. idaeus* subsp. *strigosus,* or *R. idaeus* var. *strigosus*. Recent breeding has resulted in cultivars that are thornless and more strongly upright, not needing staking.

Fruits of a golden or yellow raspberry cultivar

The black raspberry, *Rubus occidentalis,* is also cultivated, providing both fresh and frozen fruit, as well as jams, preserves, and other products, all with that species' distinctive flavor.

Purple raspberries have been produced by horticultural hybridization of red and black raspberries, and have also been found in the wild in a few places (for example, in Vermont) where the American red and the black raspberries both grow naturally. Commercial production of purple-fruited raspberries is rare.

Blue raspberry is a local name used in Prince Edward County, Ontario, Canada for the cultivar 'Columbian', a hybrid (purple raspberry) of *R. strigosus* and *R. occidentalis*.

Both the red and the black raspberry species have albino-like pale-yellow natural or horticultural variants, resulting from presence of recessive genes that impede production of anthocyanin pigments. Fruits from such plants are called golden raspberries or yellow raspberries; despite their similar appearance, they retain the distinctive flavor of their respective species (red or black). Most pale-fruited raspberries commercially sold in the eastern United States are derivatives of red raspberries. Yellow-fruited variants of the black raspberry are sometimes grown in home gardens.

Red raspberries have also been crossed with various species in other subgenera of the genus *Rubus*, resulting in a number of hybrids, the first of which was the loganberry. Later notable hybrids include boysenberry (a multi-generation hybrid), and tayberry. Hybridization between the familiar cultivated red raspberries and a few Asiatic species of *Rubus* has also been achieved.

Selected Cultivars

Numerous raspberry cultivars have been selected.

Two types of raspberry are available for commercial and domestic cultivation; the summer-bearing type produces an abundance of fruit on second-year canes (floricanes) within a relatively short period in midsummer, and double or "everbearing" plants, which also bear some fruit on first-year canes (primocanes) in the late summer and fall, as well as the summer crop on second-year canes. Those marked (AGM) have gained the Royal Horticultural Society's Award of Garden Merit.

fruits of *Rubus* 'Wyoming', a purple raspberry cultivar

a young leaf of 'Glen Prosen', a red cultivar

'Schönemann'

Red, early summer fruiting

- Boyne
- Fertödi Venus
- Rubin Bulgarski
- Cascade Dawn
- Glen Clova
- Glen Moy (AGM)
- Killarney
- Malahat
- Malling Exploit
- Malling Jewel (AGM)
- Titan
- Willamette

Red, midsummer

- Cuthbert
- Glen Ample (AGM)
- Glen Prosen (AGM)

- Lloyd George
- Meeker
- Newburgh
- Ripley
- Skeena
- Cowichan
- Chemainus
- Saanich

Red, late summer

- Cascade Delight
- Coho
- Fertödi Rubina
- Leo (AGM)
- Malling Admiral (AGM)
- Octavia
- Schoenemann
- Tulameen

Red, primocane, fall, autumn fruiting

- Amity
- Augusta
- Autumn Bliss (AGM)
- Joan J. (Thornless)
- Caroline
- Fertödi Kétszertermö
- Heritage
- Josephine
- Ripley
- Summit
- Zeva Herbsternte

Gold/Yellow, primocane, fall, autumn fruiting

- Anne
- Fallgold
- Fertödi Aranyfürt
- Goldenwest
- Golden Queen

- Honey Queen
- Kiwi Gold

Purple

- Brandywine
- Glencoe
- Royalty

Black

- Black Hawk
- Bristol
- Cumberland
- Jewel
- Munger
- Ohio Everbearer
- Scepter

Diseases and Pests

Raspberries are sometimes eaten by the larvae of some Lepidoptera species (butterflies and moths). *Botrytis cinerea*, or gray mold, is a common fungal infection of raspberries and other soft fruit under wet conditions. It is seen as a gray mold growing on the raspberries, and particularly affects fruit which are bruised, as it provides an easy entrance point for the spores.

Raspberry plants should not be planted where potatoes, tomatoes, peppers, eggplants, or bulbs have previously been grown, without prior fumigation of the soil. These crops are hosts for the disease *Verticillium* wilt, a fungus that can stay in the soil for many years and can infest the raspberry crop.

Production

In 2013, total world production of raspberries was 578,233 tonnes, with Russia supplying 25% (143,000 tonnes) (table). Other major producers were Poland (21%) and the United States (16%) (table).

Raspberry production – 2013	
Country	**Production (thousands of tonnes)**
Russia	143.0
Poland	121.0
United States	91.3

Serbia	68.5
Mexico	30.4
World	**578.2**
Source: FAOSTAT of the United Nations	

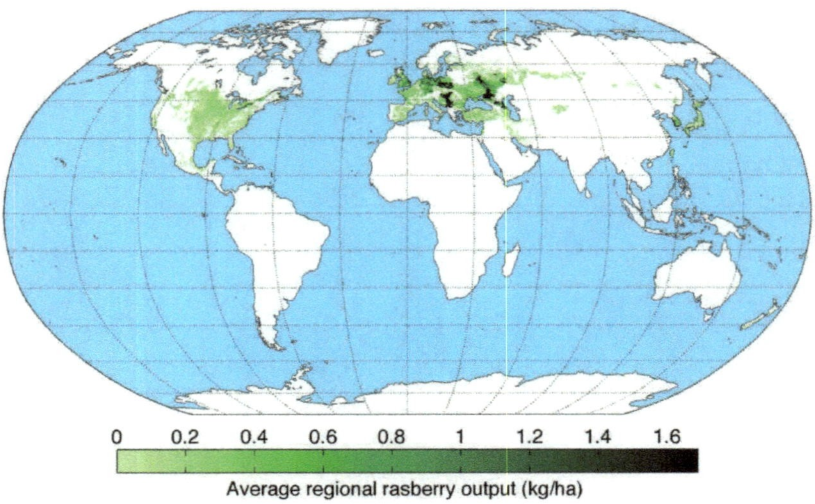

Worldwide raspberry yield

Uses

Fruit

Raspberries are grown for the fresh fruit market and for commercial processing into individually quick frozen (IQF) fruit, purée, juice, or as dried fruit used in a variety of grocery products. Traditionally, raspberries were a midsummer crop, but with new technology, cultivars, and transportation, they can now be obtained year-round. Raspberries need ample sun and water for optimal development. Raspberries thrive in well-drained soil with a pH between 6 and 7 with ample organic matter to assist in retaining water. While moisture is essential, wet and heavy soils or excess irrigation can bring on *Phytophthora* root rot, which is one of the most serious pest problems facing the red raspberry. As a cultivated plant in moist, temperate regions, it is easy to grow and has a tendency to spread unless pruned. Escaped raspberries frequently appear as garden weeds, spread by seeds found in bird droppings.

An individual raspberry weighs 3–5 g (0.11–0.18 oz), and is made up of around 100 drupelets, each of which consists of a juicy pulp and a single central seed. A raspberry bush can yield several hundred berries a year. Unlike blackberries and dewberries, a raspberry has a hollow core once it is removed from the receptacle.

Nutrients

Raspberries, raw	
Nutritional value per 100 g (3.5 oz)	
Energy	220 kJ (53 kcal)
Carbohydrates	11.94 g
Sugars	4.42 g
Dietary fiber	6.5 g
Fat	0.65 g
Protein	1.2 g
Vitamins	
Thiamine (B1)	(3%) 0.032 mg
Riboflavin (B2)	(3%) 0.038 mg
Niacin (B3)	(4%) 0.598 mg
Pantothenic acid (B5)	(7%) 0.329 mg
Vitamin B6	(4%) 0.055 mg
Folate (B9)	(5%) 21 μg
Choline	(3%) 12.3 mg
Vitamin C	(32%) 26.2 mg
Vitamin E	(6%) 0.87 mg
Vitamin K	(7%) 7.8 μg

Minerals	
Calcium	(3%) 25 mg
Iron	(5%) 0.69 mg
Magnesium	(6%) 22 mg
Manganese	(32%) 0.67 mg
Phosphorus	(4%) 29 mg
Potassium	(3%) 151 mg
Zinc	(4%) 0.42 mg
Other constituents	
Water	85.8 g
Link to USDA Database entry	
• Units • µg = micrograms • mg = milligrams • IU = International units	
Percentages are roughly approximated using US recommendations for adults. Source: USDA Nutrient Database	

Raw raspberries are 86% water, 12% carbohydrates, and have about 1% each of protein and fat (table). In a 100 gram amount, raspberries supply 53 calories.

The aggregate fruit structure contributes to raspberry's nutritional value, as it increases the proportion of dietary fiber, which is among the highest known in whole foods, up to 6% fiber per total weight. Raspberries are a rich source (20% or more of the Daily Value, DV) of vitamin C (32% DV), manganese (32% DV) and dietary fiber (26% DV) (table). Raspberries are a low-glycemic index food, with total sugar content of only 4% and no starch.

Phytochemicals

Raspberries contain phytochemicals, such as anthocyanin pigments, ellagic acid, ellagitannins, quercetin, gallic acid, cyanidins, pelargonidins, catechins, kaempferol and salicylic acid. Yellow raspberries and others with pale-colored fruits are lower in anthocyanin content. Both yellow and red raspberries contain carotenoids, mostly lutein esters, but these are masked by anthocyanins in red raspberries.

Raspberry compounds are under preliminary research for their potential to affect human health.

Leaves

Raspberry leaves can be used fresh or dried in herbal teas, providing an astringent flavor. In herbal and traditional medicine, raspberry leaves are used for some remedies, although there is no scientifically valid evidence to support their medicinal use.

BLACKBERRY

The blackberry is an edible fruit produced by many species in the *Rubus* genus in the Rosaceae family, hybrids among these species within the *Rubus* subgenus, and hybrids between the *Rubus* and *Idaeobatus* subgenera. The taxonomy of the blackberries has historically been confused because of hybridization and apomixis, so that species have often been grouped together and called species aggregates. For example, the entire subgenus *Rubus* has been called the *Rubus fruticosus* aggregate, although the species *R. fruticosus* is considered a synonym of *R. plicatus*.

Ripe, ripening, and unripe blackberries, of an unidentified blackberry species

Blackberry flower, *Rubus fruticosus*species aggregate.

Description

What distinguishes the blackberry from its raspberry relatives is whether or not the torus (receptacle or stem) "picks-with" (i.e. stays with) the fruit. When picking a blackberry fruit, the torus does stay with the fruit. With a raspberry, the torus remains on the plant, leaving a hollow core in the raspberry fruit.

The term "bramble", a word meaning any impenetrable thicket, has traditionally been applied specifically to the blackberry or its products, though in the United States it applies to all members of the *Rubus* genus. In the western US, the term caneberry is used to refer to blackberries and raspberries as a group rather than the term bramble.

The usually black fruit is not a berry in the botanical sense of the word. Botanically it is termed an aggregate fruit, composed of small drupelets. It is a widespread and well-known group of over 375 species, many of which are closely related apomictic microspecies native throughout Europe, northwestern Africa, temperate western and central Asia and North and South America.

Botanical Characteristics

Blackberries are perennial plants which typically bear biennial stems ("canes") from the perennial root system.

In its first year, a new stem, the **primocane**, grows vigorously to its full length of 3–6 m (in some cases, up to 9 m), arching or trailing along the ground and bearing large palmately compound leaves with five or seven leaflets; it does not produce any flowers. In its second year, the cane becomes a **floricane** and the stem does not grow longer, but the lateral buds break to produce flowering laterals (which have smaller leaves with three or five leaflets). First- and second-year shoots usually have numerous short-curved, very sharp prickles that are often erroneously called thorns. These prickles can tear through denim with ease and make the plant very difficult to navigate around. Prickle-free cultivars have been developed. Recently the University of Arkansas has developed primocane fruiting blackberries that grow and flower on first-year growth much as the primocane-fruiting (also called fall bearing or everbearing) red raspberries do.

Unmanaged mature plants form a tangle of dense arching stems, the branches rooting from the node tip on many species when they reach the ground. Vigorous and growing rapidly in woods, scrub, hillsides, and hedgerows, blackberry shrubs tolerate poor soils, readily colonizing wasteland, ditches, and vacant lots.

The flowers are produced in late spring and early summer on short racemes on the tips of the flowering laterals. Each flower is about 2–3 cm in diameter with five white or pale pink petals.

The drupelets only develop around ovules that are fertilized by the male gamete from a pollen grain. The most likely cause of undeveloped ovules is inadequate pollinator visits. Even a small change in conditions, such as a rainy day or a day too hot for bees to work after early morning, can reduce the number of bee visits to the flower, thus reducing the quality of the fruit. Incomplete drupelet development can also be a symptom of exhausted reserves in the plant's roots or infection with a virus such as Raspberry bushy dwarf virus.

Ecology

A bee, *Bombus hypnorum*, pollinating blackberries

Blackberry leaves are food for certain caterpillars; some grazing mammals, especially deer, are also very fond of the leaves. Caterpillars of the concealer moth *Alabonia geoffrella* have been found feeding inside dead blackberry shoots. When mature, the berries are eaten and their seeds dispersed by several mammals, such as the red fox and the Eurasian badger, as well as by small birds.

A basket of wild blackberries

Blackberries grow wild throughout most of Europe. They are an important element in the ecology of many countries, and harvesting the berries is a popular pastime. However, the plants are also considered a weed, sending down roots from branches that touch the ground, and sending up suckers from the roots. In some parts of the world without native blackberries, such as in Australia, Chile, New Zealand, and the Pacific Northwest of North America, some blackberry species, particularly *Rubus armeniacus* (Himalayan Blackberry) and *Rubus laciniatus* (Evergreen Blackberry), are naturalised and considered an invasive species and a serious weed.

Blackberry fruit are red before they are ripe, leading to an old expression that "blackberries are red when they're green".

In various parts of the United States, wild blackberries are sometimes called "Black-caps", a term more commonly used for black raspberries, *Rubus occidentalis*.

As there is forensic evidence from the Iron Age Haraldskær Woman that she consumed blackberries some 2500 years ago, it is reasonable to conclude that blackberries have been eaten by humans over thousands of years.

Cultivation

Commercial Cultivation

Worldwide, Mexico is the leading producer of blackberries, with nearly the entire crop being produced for export into the off-season fresh markets in North America and Europe. The Mexican market is almost entirely from the cultivar 'Tupy' (often spelled 'Tupi', but the EMBRAPA program in Brazil from which it was released prefers the 'Tupy' spelling). In the US, Oregon is the leading commercial blackberry producer, producing 42.6 million pounds on 6,180 acres (25.0 km^2), in 1995 and 56.1 million pounds on 7,000 acres (28 km^2) in 2009.

Black Butte blackberry

Numerous cultivars have been selected for commercial and amateur cultivation in Europe and the United States. Since the many species form hybrids easily, there are numerous cultivars with more than one species in their ancestry.

'Marion' (marketed as "marionberry") is an important cultivar that was selected from seedlings from a cross between 'Chehalem' and 'Olallie' (commonly called "Olallieberry") berries. 'Olallie' in turn is a cross between loganberry and youngberry. 'Marion', 'Chehalem' and 'Olallie' are just three of many trailing blackberry cultivars developed by the United States Department of Agriculture Agricultural Research Service (USDA-ARS) blackberry breeding program at Oregon State University in Corvallis, Oregon.

The most recent cultivars released from this program are the prickle-free cultivars 'Black Diamond', 'Black Pearl', and 'Nightfall' as well as the very early-ripening 'Obsidian' and 'Metolius'. 'Black Diamond' is now the leading cultivar being planted in the Pacific Northwest. Some of the other cultivars from this program are 'Newberry', 'Waldo', 'Siskiyou', 'Black Butte', 'Kotata', 'Pacific', and 'Cascade'.

Trailing blackberries are vigorous and crown forming, require a trellis for support, and are less cold hardy than the erect or semi-erect blackberries. In addition to the United States's Pacific Northwest, these types do well in similar climates such as the United Kingdom, New Zealand, Chile, and the Mediterranean countries.

Semi-erect, prickle-free blackberries were first developed at the John Innes Centre in Norwich, UK, and subsequently by the USDA-ARS in Beltsville, Maryland. These are crown forming and very vigorous and need a trellis for support. Cultivars include 'Black Satin' 'Chester Thornless', 'Dirksen Thornless', 'Hull Thornless', 'Loch Maree', 'Loch Ness', 'Loch Tay', 'Merton Thornless', 'Smoothstem', and 'Triple Crown'. Recently, the cultivar 'Cacanska Bestrna' (also called 'Cacak Thornless') has been developed in Serbia and has been planted on many thousands of hectares there.

The University of Arkansas has developed cultivars of erect blackberries. These types are less vigorous than the semi-erect types and produce new canes from root initials (therefore they spread underground like raspberries). There are prickly and prickle-free cultivars from this program, including 'Navaho', 'Ouachita', 'Cherokee', 'Apache', 'Arapaho', and 'Kiowa'. They are also responsible for developing the primocane fruiting blackberries such as 'Prime-Jan' and 'Prime-Jim'.

In raspberries, these types are called primocane fruiting, fall fruiting, or everbearing. 'Prime-Jim' and 'Prime-Jan' were released in 2004 by the University of Arkansas and are the first cultivars of primocane fruiting blackberry. They grow much like the other erect cultivars described above, however the canes that emerge in the spring, will flower in mid-summer and fruit in late summer or fall. The fall crop has its highest quality when it ripens in cool mild climate such as in California or the Pacific Northwest.

'Illini Hardy' a semi-erect prickly cultivar introduced by the University of Illinois is cane hardy in zone 5, where traditionally blackberry production has been problematic, since canes often failed to survive the winter.

Blackberry production in Mexico has expanded enormously in the past decade. While once based on the cultivar 'Brazos', an old erect blackberry cultivar developed in Texas in 1959, the Mexican industry is now dominated by the Brazilian 'Tupy' released in the 1990s. 'Tupy' has the erect blackberry 'Comanche', and a "wild Uruguayan blackberry" as parents. Since there are no native blackberries in Uruguay, the suspicion is that the widely grown 'Boysenberry' is the male parent. In order to produce these blackberries in regions of Mexico where there is no winter chilling to stimulate flower bud development, chemical defoliation and application of growth regulators are used to bring the plants into bloom.

Diseases and Pests

As a result of blackberries belonging to the same genus as raspberries, they share the same diseases including anthracnose which can cause the berry to have uneven ripening and sap flow may also be slowed. They also share the same remedies including the Bordeaux mixture, a combination of lime, water and copper(II) sulfate. The rows between blackberry plants must be free of weeds, blackberry suckers and grasses which may lead to pests or diseases. Fruit growers are selective when planting blackberry bushes as wild blackberries may be infected and gardeners are recommended to purchase only certified disease-free plants.

Raindrop on blackberry pale pink flower

The spotted-wing drosophila, *Drosophila suzukii* is a serious pest of blackberries. Unlike its vinegar fly relatives which are primarily attracted to rotting or fermented fruit, *D. suzukii* attacks fresh, ripe fruit by laying eggs under the soft skin. The larvae hatch and grow in the fruit, destroying the fruit's commercial value.

Another pest is *Amphorophora rubi*, known as the Blackberry Aphid, which not only eats blackberries but raspberries as well.

Byturus tomentosus (Raspberry beetle), *Lampronia corticella* (Raspberry Moth) and *Anthonomus rubi* (Strawberry blossom weevil) are also known to infest blackberries.

STRAWBERRY

The garden strawberry (or simply strawberry; *Fragaria* × *ananassa*) is a widely grown hybrid species of the genus *Fragaria* (collectively known as the strawberries). It is cultivated worldwide for its fruit. The fruit (which is not a botanical berry, but an aggregate accessory fruit) is widely appreciated for its characteristic aroma, bright red color, juicy texture, and sweetness. It is consumed in large quantities, either fresh or in such prepared foods as preserves, fruit juice, pies, ice creams, milkshakes, and chocolates. Artificial strawberry flavorings and aromas are also widely used in many products like lip gloss, candy, hand sanitizers, perfume, and many others.

The garden strawberry was first bred in Brittany, France, in the 1750s via a cross of *Fragaria virginiana* from eastern North America and *Fragaria chiloensis*, which was brought from Chile by Amédée-François Frézier in 1714. Cultivars of *Fragaria* × *ananassa* have replaced, in commercial production, the woodland strawberry (*Fragaria vesca*), which was the first strawberry species cultivated in the early 17th century.

Strawberry fruit

Strawberry fruit cross section

Technically, the strawberry is an aggregate accessory fruit, meaning that the fleshy part is derived not from the plant's ovaries but from the receptacle that holds the ovaries. Each apparent "seed" (achene) on the outside of the fruit is actually one of the ovaries of the flower, with a seed inside it.

Cultivation

Strawberry cultivars vary widely in size, color, flavor, shape, degree of fertility, season of ripening, liability to disease and constitution of plant.

Strawberry field in North Rhine-Westphalia, Germany

On average, a strawberry has about 200 seeds on its external membrane. Some vary in foliage, and some vary materially in the relative development of their sexual organs. In most cases, the flowers appear hermaphroditic in structure, but function as either male or female. For purposes of commercial production, plants are propagated from runners and, in general, distributed as either bare root plants or plugs. Cultivation follows one of two general models—annual plasticulture, or a perennial system of matted rows or mounds. Greenhouses produce a small amount of strawberries during the off season.

Fragaria × *ananassa* in the UBC Botanical Garden.

A field using the plasticulture method

Strawberry field at Bedugul, Bali, covered with plastic

The bulk of modern commercial production uses the plasticulture system. In this method, raised beds are formed each year, fumigated, and covered with plastic to prevent weed growth and erosion. Plants, usually obtained from northern nurseries, are planted through holes punched in this covering, and irrigation tubing is run underneath. Runners are removed from the plants as they appear, in order to encourage the plants to put most of their energy into fruit development. At the end of the harvest season, the plastic is removed and the plants are plowed into the ground.Because strawberry plants more than a year or two old begin to decline in productivity and fruit quality, this system of replacing the plants each year allows for improved yields and denser plantings. However, because it requires a longer growing season to allow for establishment of the plants each year, and because of the increased costs in terms of forming and covering the mounds and purchasing plants each year, it is not always practical in all areas.

The other major method, which uses the same plants from year to year growing in rows or on mounds, is most common in colder climates. It has lower investment costs, and lower overall maintenance requirements. Yields are typically lower than in plasticulture.

Another method uses a compost sock. Plants grown in compost socks have been shown to produce significantly higher oxygen radical absorbance capacity (ORAC), flavonoids, anthocyanins, fructose, glucose, sucrose, malic acid, and citric acid than fruit produced in the black plastic mulch or matted row systems. Similar results in an earlier 2003 study conducted by the US Dept of Agriculture, at the Agricultural Research Service, in Beltsville Maryland, confirms how compost plays a role in the bioactive qualities of two strawberry cultivars.

Strawberries are often grouped according to their flowering habit. Traditionally, this has consisted of a division between "June-bearing" strawberries, which bear their fruit in the early summer and "ever-bearing" strawberries, which often bear several crops of fruit throughout the season. Research published in 2001 showed that strawberries actually occur in three basic flowering habits: short-day, long-day, and day-neutral. These refer to the day-length sensitivity of the plant and the type of photoperiod that induces flower formation. Day-neutral cultivars produce flowers regardless of the photoperiod.

Strawberries may also be propagated by seed, though this is primarily a hobby activity, and is not widely practiced commercially. A few seed-propagated cultivars have been developed for home use, and research into growing from seed commercially is ongoing. Seeds (achenes) are acquired either via commercial seed suppliers, or by collecting and saving them from the fruit.

Strawberries can also be grown indoors in strawberry pots.

Kashubian strawberry (*Truskawka kaszubska* or *Kaszëbskô malëna*) are the first Polish fruit to be given commercial protection under EU law. They are produced in Kartuzy, Kościerzyna and Bytów counties and in the municipalities of Przywidz, Wejherowo, Luzino, Szemud, Linia, Łęczyce and Cewice in Kashubia. Only the following varieties may be sold as Senga Sengana, Elsanta, Honeoye that have been graded as Extra or Class I.

Manuring and Harvesting

Most strawberry plants are now fed with artificial fertilizers, both before and after harvesting, and often before planting in plasticulture.

To maintain top quality, berries are harvested at least every other day. The berries are picked with the caps still attached and with at least half an inch of stem left. Strawberries need to remain on the plant to fully ripen because they do not continue to ripen after being picked. Rotted and overripe berries are removed to minimize insect and disease problems. The berries do not get washed until just before consumption. They are covered in a shallow pan and refrigerated when storing.

A diorama created from beeswax by Dr. Henry Brainerd Wright at the Louisiana State Exhibit Museum in Shreveport, Louisiana depicts strawberry harvesting. Strawberries are particularly grown in the southeastern portion of the state around Hammond.

Harvest

Soil test information and plant analysis results are used to determine fertility practices. Nitrogen fertilizer is needed at the beginning of every planting year. There are normally adequate levels of phosphorus and potash when fields have been fertilized for top yields. In order to provide more organic matter a cover crop of wheat or rye is planted in the winter the year before planting the strawberries. Strawberries prefer a pH from 5.5 to 6.5 so lime is usually not applied.

The harvesting and cleaning process has not changed substantially over time. The delicate strawberries are still harvested by hand. Grading and packing often occurs in the field, rather than in a processing facility. In large operations, strawberries are cleaned by means of water streams and shaking conveyor belts.

Pests

Around 200 species of pests are known to attack strawberries both directly and indirectly. These pests include slugs, moths, fruit flies, chafers, strawberry root weevils, strawberry thrips, strawberry sap beetles, strawberry crown moth, mites, aphids, and others. The caterpillars of a number of species of Lepidoptera feed on strawberry plants.

The strawberry aphid, *Chaetosiphon fragaefolii*, is a bug species found in the United States (Arizona), Argentina and Chile. It is a vector of the strawberry mild yellow-edge virus.

The amounts of pesticides required for industrial production of strawberries (300 pounds (140 kg) in California per acre) have led to the strawberry leading the list of EWG's "Dirty Dozen" of pesticide-contaminated produce.

Diseases

Strawberry plants can fall victim to a number of diseases. The leaves may be infected by powdery mildew, leaf spot (caused by the fungus *Sphaerella fragariae*), leaf blight (caused by the fungus *Phomopsis obscurans*), and by a variety of slime molds. The crown and roots may fall victim to red stele, verticillium wilt, black root rot, and nematodes. The fruits are subject to damage from gray mold, rhizopus rot, and leather rot. To prevent root-rotting, strawberries should be planted every four to five years in a new bed, at a different site.

The plants can also develop disease from temperature extremes during winter. When watering strawberries, advice has been given to water only the roots and not the leaves, as moisture on the leaves encourages growth of fungus.

Production Trends

World strawberry production in tons							
Rank	**Country**	**2006**	**2007**	**2008**	**2009**	**2010**	**2011**
1	USA	1,090,436	1,109,215	1,148,350	1,270,640	1,294,180	1,312,960
2	Turkey	211,127	250,316	261,078	291,996	299,940	302,416
3	Spain	330,485	269,139	281,240	266,772	275,355	262,730
4	Egypt	128,349	174,414	200,254	242,776	238,432	240,284

5	Mexico	191,843	176,396	207,485	233,041	226,657	228,900
6	Russia	227,000	230,400	180,000	185,000	165,000	184,000
7	Japan	190,700	191,400	190,700	184,700	177,500	177,300
8	South Korea	205,307	203,227	192,296	203,772	231,803	171,519
9	Poland	193,666	174,578	200,723	198,907	153,410	166,159
10	Germany	173,230	158,658	150,854	158,563	156,911	154,418
11	Italy	143,315	160,558	155,583	163,044	153,875	150,000
	Total world	3,973,243	4,001,721	4,136,802	4,596,614	4,366,889	4,594,539

Strawberries for sale at Mahabaleshwar

Although official numbers are not available, the strawberry production of China in market year 2011/12 is estimated to be 2,100,000 tons.

Domestic Cultivation

Strawberries are popular and rewarding plants to grow in the domestic environment, be it for consumption or exhibition purposes, almost anywhere in the world. The best time to plant is in late summer or spring. Plant in full sun or dappled shade, and in somewhat sandy soil. The addition of manure and a balanced fertilizer aids strong growth. Alternatively they can be planted in pots or special planters using compost. Fibre mats placed under each plant will protect fruits from touching the ground, and will act as a weed barrier.

Garden strawberry flower

Picking home-grown garden strawberries

Strawberries are tough and will survive many conditions, but during fruit formation, moisture is vital, especially if growing in containers. Moreover, protection must be provided against slugs and snails which attack the ripe fruit. The fruit matures in midsummer and should be picked when fully ripe — that is, the fruit is a uniform bright red color. The selection of different varieties can extend the season in both directions. Numerous cultivars have been selected for consumption and for exhibition purposes. The following cultivars have gained the Royal Horticultural Society's Award of Garden Merit:-

• 'Cambridge Favourite' • 'Hapil'	• 'Honeoye' • 'Pegasus'	• 'Rhapsody' • 'Symphony'

Propagation is by runners, which can be pegged down to encourage them to take root, or cut off and placed in a new location. Established plants should be replaced every three years, or sooner if there are signs of disease.

When propagating strawberries, one should avoid using the same soil or containers that were previously used for strawberry cultivation. After cultivating strawberries, rotating to another culture is advisable, because diseases that attack one species might not attack another.

REFERENCES

1. "Define Etaerio at Dictionary.com". Retrieved 2015-01-17.
2. Hickey, M.; King, C. (2001). The Cambridge Illustrated Glossary of Botanical Terms. Cambridge University Press.
3. Spjut, R.; Thieret, J. (1989). "Confusion between multiple and aggregate fruits". The Botanical Review. 55 (1): 53–72. doi:10.1007/bf02868781.
4. Beentje, H.; Williamson, J. (2010). The Kew Plant Glossary: an Illustrated Dictionary of Plant Terms. Royal Botanic Gardens, Kew: Kew Publishing.
5. Janick, J.; Paull, R.E. (2008). The Encyclopedia of Fruit and Nuts. CABI. ISBN 9780851996387.

CHAPTER 7

Cultivation of Multiple Fruit

Chapter Contents

A multiple fruit or composite fruit develops from an entire inflorescence i.e., from a cluster or group of flowers into an entire fruit. Each of the flower produces a fruit, but these all mature into a single mass to form the composite fruit. In simple words we can say that - Fruits of individual flowers closely packed to form a single fruit.

INTRODUCTION

Multiple fruits, also called collective fruits, are fruiting bodies formed from a cluster of fruiting flowers, the *inflorescence*. Each flower in the inflorescence produces a fruit, but these mature into a single mass in which each flower has produced a true fruit. After flowering the mass is called an infructescence. Examples are the fig, pineapple, mulberry, osage-orange, and breadfruit.

In languages other than English, the meanings of multiple and aggregate fruit are reversed, so that multiple fruits merge several pistils within a single flower.

In some plants, such as this noni, flowers are produced regularly along the stem and it is possible to see together examples of flowering, fruit development, and fruit ripening

In the photograph on the left, stages of flowering and fruit development in the noni or Indian mulberry (*Morinda citrifolia*) can be observed on a single branch. First an inflorescence of white flowers called a head is produced. After fertilization, each flower develops into a drupe, and as the drupes expand, they become *connate* (merge) into a *multiple fleshy fruit* called a *syncarp*. There are also many dry multiple fruits.

Other examples of multiple fruits:

- Platanus, multiple achenes from multiple flowers, in a single fruit structure
- Mulberry, multiple flowers form one fruit
- Fig, multiple flowers similar to mulberry infructescence form a multiple fruit inside the inverted inflorescence. This form is called a Syconium

MORUS

Morus, a genus of flowering plants in the family Moraceae, comprises 10–16 species of deciduous trees commonly known as mulberries, growing wild and under cultivation in many temperate world regions.

The closely related genus *Broussonetia* is also commonly known as mulberry, notably the paper mulberry, *Broussonetia papyrifera*. Mulberries are fast-growing when young, but soon become slow-growing and rarely exceed 10–15 metres (30–50 ft) tall. The leaves are alternately arranged, simple and often lobed and serrated on the margin. Lobes are more common on juvenile shoots than on mature trees.

The trees can be monoecious or dioecious. The mulberry fruit is a multiple fruit, approximately 2 to 3 cm (one inch) long. Immature fruits are white, green, or pale yellow. In most species the fruits turn pink and then red while ripening, then dark purple or black, and have a sweet flavor when fully ripe. The fruits of the white-fruited cultivar are white when ripe; the fruit of this cultivar is also sweet, but has a very bland flavor compared with darker varieties.

Morus nigra

Distribution and Cultivation

Black, red, and white mulberry are widespread in southern Europe, the Middle East, northern Africa and Indian subcontinent, where the tree and the fruit have names under regional dialects. Jams and sherbets are often made from the fruit in this region. Black mulberry was imported to Britain in the 17th century in the hope that it would be useful in the cultivation of silkworms. It was much used in folk medicine, especially in the treatment of ringworm. Mulberries are also widespread in Greece, particularly in the Peloponnese, which in the Middle Ages was known as Morea, deriving from the Greek word for the tree (μουριά, *mouria*).

Mulberry fruit in Libya

Mulberries can be grown from seed, and this is often advised as seedling-grown trees are generally of better shape and health, but they are most often planted from large cuttings which root readily. The mulberry plants which are allowed to grow tall with a crown height of 5–6 feet from ground level and a stem girth of 4–5 inches or more is called tree mulberry. They are specially raised with the help of well-grown saplings 8–10 months old of any of the varieties recommended for rain-fed areas like S-13 (for red loamy soil) or S-34 (black cotton soil) which are tolerant to drought or soil-moisture stress conditions. Usually, the plantation is raised and in block formation with a spacing of 6 feet × 6 feet, or 8 feet × 8 feet, as plant to plant and row to row distance. The plants are usually pruned once a year during the monsoon season (July – August) to a height of 5–6 feet and allowed to grow with a maximum of 8–10 shoots at the crown. The leaves are harvested three or four times a year by a leaf-picking method under rain-fed or semiarid conditions, depending on the monsoon.

The tree branches pruned during the fall season (after the leaves have fallen) are cut and used to make durable baskets supporting agriculture and animal husbandry.

Some North American cities have banned the planting of mulberries because of the large amounts of pollen they produce, posing a potential health hazard for some pollen allergy sufferers. In actuality, only the male mulberry trees produce pollen; this light-weight pollen can be inhaled deeply into the lungs, sometimes triggering asthma. Conversely, female mulberry trees produce all-female flowers, which draw pollen and dust from the air. Because of this pollen-absorbing feature, all-female mulberry trees have an OPALS allergy scale rating of just 1 (lowest level of allergy potential), and some consider it "allergy-free".

Fortunately, mulberry tree scion wood can easily be grafted onto other mulberry trees during the winter, when the tree is dormant. One common scenario is converting a problematic male mulberry tree to an allergy-free female tree, by grafting all-female mulberry tree scions to a male mulberry that has been pruned back hard. However, any new growth from below the graft(s) must be removed, as they would be from the original male mulberry tree.

FICUS

Ficus is a genus of about 850 species of woody trees, shrubs, vines, epiphytes and hemiepiphytes in the family Moraceae. Collectively known as fig trees or figs, they are native throughout the tropics with a few species extending into the semi-warm temperate zone. The common fig (*F. carica*) is a temperate species native to southwest Asia and the Mediterranean region (from Afghanistan to Portugal), which has been widely cultivated from ancient times for its fruit, also referred to as figs. The fruit of most other species are also edible though they are usually of only local economic importance or eaten as bushfood. However, they are extremely important food resources for wildlife. Figs are also of considerable cultural importance throughout the tropics, both as objects of worship and for their many practical uses.

Sycamore fig, *Ficus sycomorus*

Production

Top five producers of raw figs (2013, in millions of tonnes)	
Turkey	0.3
Egypt	0.15
Algeria	0.12
Morocco	0.1
Iran	0.08
World	**1.1**
Source: United Nations FAOSTAT	

World production of raw figs in 2013 was 1.1 million tonnes, led by Turkey with 0.3 million tonnes, double the volume of the next largest producer, Egypt (table). Dried fig production in 2013 was 133,447 tonnes globally, also led by Turkey with 36% of the total, followed by Morocco and Iran.

Fig Fruit and Reproduction System

Many fig species are grown for their fruits, though only *Ficus carica* is cultivated to any extent for this purpose. The fig fruits, important as both food and traditional medicine, contain laxative substances, flavonoids, sugars, vitamins A and C, acids and enzymes. However, figs are skin allergens, and the latex is a serious eye irritant.

Ficus exasperata, fruits

A fig "fruit" is a type of multiple fruit known as a syconium, derived from an arrangement of many small flowers on an inverted, nearly closed receptacle. The many small flowers are unseen unless the fig is cut open. In Chinese the fig is called *wú hu guo*, "fruit without flower". In Bengali, where the common fig is called *dumur*, it is referenced in a proverb: *tumi jeno dumurer phool hoe gele* ("You have become [invisible like] the *dumur* flower").

The syconium often has a bulbous shape with a small opening (the ostiole) at the outward end that allows access to pollinators. The flowers are pollinated by very small wasps that crawl through the opening in search of a suitable place to lay eggs. Without this pollinator service fig trees could not reproduce by seed. In turn, the flowers provide a safe haven and nourishment for the next generation of wasps. This accounts for the frequent presence of wasp larvae in the fruit, and has led to a coevolutionary relationship. Technically, a fig fruit proper would be only one of the many tiny matured, seed-bearing gynoecia found inside one fig – if you cut open a fresh fig, individual fruit will appear as fleshy "threads", each bearing a single seed inside. The genus *Dorstenia*, also in the fig family (Moraceae), exhibits similar tiny flowers arranged on a receptacle but in this case the receptacle is a more or less flat, open surface.

Fig plants can be monoecious (hermaphrodite) or gynodioecious (hermaphrodite and female). Nearly half of fig species are gynodioecious, and therefore have some plants with inflorescences (syconium) with long styled pistillate flowers, and other plants with staminate flowers mixed with short styled pistillate flowers. The long flowers styles tend to prevent wasps from laying their eggs within the ovules, while the short styled flowers are accessible for egg laying.

All the native fig trees of the American continent are hermaphrodites, as well as species like Indian banyan (*F. benghalensis*), weeping fig (*F. benjamina*), Indian rubber plant (*F. elastica*), fiddle-leaved fig (*F. lyrata*), Moreton Bay fig (*F. macrophylla*), Chinese banyan (*F. microcarpa*), sacred Fig (*F. religiosa*) and sycamore fig (*F. sycomorus*).

On the other hand, the common fig (*Ficus carica*) is a gynodioecious plant, as well as lofty fig or clown fig (*F. aspera*), Roxburgh fig (*F. auriculata*), mistletoe fig (*F. deltoidea*), *F. pseudopalma*, creeping fig (*F. pumila*) and related species.

The hermaphrodite common figs are called "inedible figs" or caprifigs; in traditional culture in the Mediterranean region they were considered food for goats (*Capra aegagrus*). In the female fig trees, the male flower parts fail to develop; they produce the "edible figs". Fig wasps grow in common fig caprifigs but not in the female syconiums because the female flower is too long for the wasp to successfully lay her eggs in them. Nonetheless, the wasp pollinates the flower with pollen from the caprifig it grew up in. When the wasp dies, it is broken down by enzymes (Ficain) inside the fig. Fig wasps are not known to transmit any diseases harmful to humans.

When a caprifig ripens, another caprifig must be ready to be pollinated. In temperate climes, wasps hibernate in figs, and there are distinct crops. Common fig caprifigs have three crops per year; edible figs have two. The first (breba) produces small fruits called olynth. Some parthenocarpic cultivars of common figs do not require pollination at all, and will produce a crop of figs (albeit sterile) in the absence of caprifigs or fig wasps.

Depending on the species, each fruit can contain hundreds or even thousand of seeds. Figs can be propagated by seeds, cuttings, air-layering or grafting. However, as with any plant, figs grown from seed are not necessarily genetically identical to the parent and are only propagated this way for breeding purposes.

Mutualism with the Pollinating Fig Wasps

There is typically only one species of wasp capable of fertilizing the flowers of each species of fig, and therefore plantings of fig species outside of their native range results in effectively sterile individuals. For example, in Hawaii, some 60 species of figs have been introduced, but only four of the wasps that fertilize them have been introduced, so only four species of figs produce viable seeds there and can become invasive species. This is an example of mutualism, in which each organism (fig plant and fig wasp) benefit each other, in this case reproductively.

The intimate association between fig species and their wasp pollinators, along with the high incidence of a one-to-one plant-pollinator ratio have long led scientists to believe that figs and wasps are a clear example of coevolution. Morphological and reproductive behavior evidence, such as the correspondence between fig and wasp larvae maturation rates, have been cited as support for this hypothesis for many years. Additionally, recent genetic and molecular dating analyses have shown a very close correspondence in the character evolution and speciation phylogenies of these two clades.

Recently, molecular techniques including the combined use of microsatellite markers in combination with mitochondrial sequence analyses have suggested that the one-to-one relationships between figs and their pollinators may not be as strict as once believed The discovery of multiple genetically distinct, cryptic wasp species paired with individual fig species supports this concern, particularly considering that not all cryptic species are sister taxa and thus must have experienced a host shift at some point. These cryptic species lacked evidence of genetic introgression or backcrosses indicating limited fitness for hybrids and effective reproductive isolation and speciation.

The existence of cryptic species suggests that neither the number of symbionts nor their evolutionary relationships are necessarily fixed ecologically. Fifty percent of fig species host multiple wasp pollinators thus are not tied inextricably to any single symbiont. On the other hand, species of wasps have been shown to pollinate multiple host fig species While the morphological characteristics that facilitate the fig-wasp mutualisms are likely to be shared more fully in closer relatives, the absence of unique pairings would make it impossible to do a one-to-one tree comparison and difficult to determine cospeciation.

Systematics

With 800 species, *Ficus* is by far the largest genus in the Moraceae, and is one of the largest genera of flowering plants currently described. The species currently classified within *Ficus* were originally split into several genera in the mid-1800s, providing the basis for a subgeneric classification when reunited into one genus in 1867. This classification put functionally dioecious species into four subgenera based on floral characters. In 1965, E. J. H. Corner reorganized the genus on the basis of breeding system, uniting these four dioecious subgenera into a single dioecious subgenus *Ficus*. Monoecious figs were classified within the subgenera *Urostigma*, *Pharmacosycea* and *Sycomorus*.

This traditional classification been called into question by recent phylogenetic studies employing genetic methods to investigate the relationships between representative members of the various sections of each subgenus. Of Corner's original subgeneric divisions of the genus, only *Sycomorus* is supported as monophyletic in the majority of phylogenetic studies. Notably, there is no clear split between dioecious and monoecious lineages. One of the two sections of *Pharmacosycea*, a monoecious group, form a monophyletic clade basal to the rest of the genus, which includes the other section of *Pharmacosycea*, the rest of the monoecious species, and all of the dioecious species. These remaining species are divided into two main monophyletic lineages (though the statistical support for these lineages isn't as strong as for the monophyly of the more derived clades within them). One consists of all sections of *Urostigma* except for section *Urostigma s. s.*. The other includes section *Urostigma s. s.*, subgenus *Sycomorus*, and the species of subgenus *Ficus*, though the relationships of the sections of these groups to one another are not well resolved.

REFERENCES

1. Schlegel. Encyclopedic Dictionary of Plant Breeding and Related Subjects. p. 282.
2. Hickey, M.; King, C. (2001). The Cambridge Illustrated Glossary of Botanical Terms. Cambridge University Press.
3. Beentje, H.; Williamson, J. (2010). The Kew Plant Glossary: an Illustrated Dictionary of Plant Terms. Royal Botanic Gardens, Kew: Kew Publishing.
4. Spjut, R.; Thieret, J. (1989). "Confusion between multiple and aggregate fruits". The Botanical Review. 55 (1): 53–72. doi:10.1007/bf02868781.

Index

O

P

R

S

T

U

V